Living with Tomorrow

Living with Tomorrow

A Factual Look at America's Resources

RICHARD M. STEPHENSON
Department of Chemical Engineering
University of Connecticut

A Wiley-Interscience Publication
JOHN WILEY & SONS
New York • Chichester • Brisbane • Toronto • Singapore

Library of Congress Cataloging in Publication Data:

Stephenson, Richard Montgomery, 1917-
Living with tomorrow.

"A Wiley-Interscience publication."
Bibliography: p.
Includes index.
1. Power resources—United States. 2. Natural resources—United States. 3. Ecology—United States. I. Title.

TJ163.25.U6S76 333.79'0973 81-10467
ISBN 0-471-09457-9 AACR2

Printed in the United States of America

10 9 8 7 6 5 4 3 2 1

Preface

Since the rude awakening of the 1973–74 energy crisis, our American way of life has been increasingly threatened by questions of energy, food, raw materials, pollution, and human resources. It has been further threatened by a lack of perspective on many levels of authority and by a common inability to view the choices we face without distortion or exaggeration. Two facts stand out at this point: United States resources have limits, and demands on these resources are presently frozen by law as well as by tradition into a pattern of growth. Under these circumstances, it is not surprising to find growing concern about the future of our country and whether, in fact, our system of government can survive in a resource-limited world.

Even further complicating the situation is the need for a certain amount of technical background to comprehend the problems before us. No longer can such disciplines as economics, political science, sociology, and technology be treated as separate and distinct fields of study. Today's overriding problems literally cut across all of these fields. Energy and pollution are just as important to the political scientist as to the engineer; unemployment and productivity concern the sociologist as much as the economist. Without a basic understanding of these areas, it will be impossible for anyone to function intelligently in the two final decades of the twentieth century.

In providing an evenhanded, factual presentation of the choices we face on questions of population, food, energy, raw materials, pollution, human resources, and economic survival in a world of diminishing resources, together with an explanation of certain related mathematical growth laws, *Living with Tomorrow* seeks to give nontechnical readers some much-needed tools for arriving at sound conclusions of their own. For this reason, moralizing and prescriptive material are kept to a minimum. Intended primarily for the general public, this book is written in simple yet scientifically accurate language and will also serve as a supplementary text for classes in economics, sociology, political science, energy, and environmental policy at four-year and two-year colleges and in the more sophisticated high school programs.

I would like to thank Tom McQueen for preparing the final drawings for publication, John Hall for his help with the photographs, and the staff of the University of Connecticut library for their assistance with interlibrary loans and government documents. Above all, I must thank my wife Mary who has served (among other things) as copy editor, typist, and devil's advocate, and who tries to keep my feet on the ground when I sound too much like a college professor.

RICHARD M. STEPHENSON

Storrs, Connecticut
August 1981

Contents

Chapter One

An Introduction to the Real World

Saturday, October 6, 1973, was a cool and cloudy day in the Middle East. In Israel, it was a doubly holy day: Saturday was the sabbath, and October 6 was Yom Kippur, the Day of Atonement and most solemn holy day of the Hebrew calendar. The people were either at home or in their synagogues; schools and businesses were closed; there was no public transportation, no traffic in the streets, no public entertainment, and no radio or television. Many Israeli soldiers were home on leave for Yom Kippur. The Bar-Lev defense line along the Suez Canal was manned by fewer that 600 men, mostly middle-aged reservists, many in their last tour of duty before retirement. Two armored brigades were stationed in the Golan area along the Syrian border, their tanks unmanned as crews chatted beside them; some of the more religious were reciting Yom Kippur prayers.

By coincidence, October 6 also had a special significance for the Arabs. On that day in the year 624, the prophet Muhammad began preparations for the Battle of Badr, the first victory in a long campaign, which ended in 630 with his triumphant entry into Mecca and marked the start of the spread of Islam. For several years, the Arabs had been planning a new war on Israel. The military assault was code-named *Operation BADR,* and October 6, 1973, was D day.

Figure 1. Going to work the American way: rush-hour traffic on a Los Angeles freeway. (Richard Allman photo).

Shortly after noon, artillery shells began flying from Egyptian batteries along the Suez Canal. This was not surprising, since periodic bombardments were a way of life along the Canal. This time, however, canisters belching dark green smoke set up smoke screens, and Egyptian commando boats were soon making their way across the Canal. At 1:58 p.m. five Syrian MiG-17 fighters skimmed low to strafe Israeli positions in the Golan area. This was followed by an artillery attack as 700 Syrian tanks started to roll across the Golan "cease-fire" strip. The fourth Arab-Israeli war was under way!

Meanwhile, for Israelis at home, the first indication of trouble was an air raid warning at 2 p.m., when sirens started to wail over every city and hamlet of the nation. At 2:40 p.m., Israeli radio came back on the air, announced the war, and began calling reserves by code name for units being mobilized: "Sea wolf, Sea wolf . . . Hound tooth, Hound tooth . . . Desert rat, Desert rat . . ." Couriers ran to homes and synagogues with lists of names being called, rabbis hastily blessed food for the soldiers (Yom Kippur is a day of fasting), and reservists made their way to the battlefields by any means they could muster. Private cars were commandeered; aging buses, bread vans, and even garbage trucks were soon carrying men to the front. Many hitched rides, some still wearing their prayer caps.

Sixteen days later, this war ended as the result of a United Nations cease-fire mandate, which was strongly supported by both the United States and the Soviet Union. Both Israelis and Arabs had suffered heavy losses of men and materiel. Neither side had achieved a clear-cut military victory. The final peace settlement gave Israel a small amount of additional territory in the Golan area and some land on the west bank of the Suez Canal. Egypt gained some territory on the east bank of the Canal. The Israelis acquired more respect for the Arab fighting man. The great powers realized once more the dangers inherent in this volatile part of the world and the necessity for achieving a stable and lasting peace. Only now is it becoming obvious to us that those few days also marked the end of cheap energy for the United States and the western world—and very possibly the end of an economic growth curve, a way of life we have known for some 200 years.

A GLANCE BACKWARD

Until the last half of the eighteenth century, all of the nations of the world existed as largely rural societies. There was no industry as we know it. Most of the population of England and the United States was engaged in farming of one kind or another, with their simple needs supplied by family-owned shops and manufactories, local foundries, the village blacksmith, and other domestic sources. Machinery had changed little over the past hundred years, transportation was very limited, and almost all individuals were entirely dependent on food and materials produced in their immediate neighborhood.

During the final decades of the eighteenth century, however, certain mechanical inventions in England revolutionized this domestic industry, particularly the manufacture of cotton. In 1769, the water frame was patented by Arkwright, with the spinning jenny by Hargreaves in the following year. Alone, these would not have been revolutionary, but 1769 was also the year James Watt took out his patent for the steam engine, and the power loom was patented by Cartwright in 1785. The stage was set for the industrial revolution of the nineteenth century.

No longer dependent on water power, large manufacturing plants mushroomed throughout England. Coke replaced charcoal for the smelting of iron, the steam engine supplied compressed air for the blast furnace, and the coal and iron industries expanded rapidly to supply the new industrial demands. Country people poured into the cities, and many pleasant villages became overcrowded cities almost overnight. In 1800, there were twelve cities in England with a population over 20,000; by 1900, there were nearly 200.

Changes in the United States were even more dramatic. Industrial production was stimulated by the American Revolution, and mass production of standardized parts spread rapidly with the Civil War. In 1800, the population of the United States was only 5.3 million, with only three of its cities (Boston, New York, and Philadelphia) over 25,000; by 1850, the total population was 23 million; and by 1900 this had grown to 75 million, with thirty-eight cities over 100,000.

Despite the much-publicized misery in newly crowded cities, mills, and factories, with child labor and workdays that lasted from sunup to sundown, the industrial revolution brought many benefits to workers of all classes on both sides of the Atlantic. People swarmed to the new manufacturing establishments because they wanted to work in them. By today's standards, working conditions were certainly deplorable; yet, by the standards of the time, working conditions in the mills were evidently preferable to life on the farm, pay was high enough to attract competent workers, and there was a certain excitement in the bustling cities. By the early nineteenth century, the rate of the population growth in England was sharply increasing, owing to early marriages and to notable progress in sanitary conditions, which improved the health of the working people.

The industrial revolution also brought a great increase in the production of consumer goods. The average worker had more and better food, better clothing, better medical care, and possibly even better housing than before. Each generation could now look forward to a more affluent life than its parents. Except for setbacks from wars and from periodic "hard times" the standard of living in our industrialized nations improved steadily for almost 200 years.

An even more rapid upswing in our standard of living resulted from the scientific and technological developments which began to appear toward the end of the nineteenth century. In the last fifty to seventy-five years, just one human lifetime, our way of life has changed more than in all previously recorded history. The twentieth century has brought us the automobile, airplane travel, X rays, atomic energy, radio, television, synthetic polymers, and antibiotics—to name just a few "improvements." Infectious disease has been essentially conquered; in this country today over half the deaths are cardiovascular (often stress-related), 20 percent from malignancies, and 5 percent from accidents. All other causes together account for less than one quarter of these deaths. Our greatest health hazards are now smoking and obesity.

TOO MUCH, TOO SOON?

The past forty years have seen a tremendous explosion of knowledge—if not of wisdom. Some 90 percent of all the scientists who have ever lived are alive today. About 80 percent of the population of the United States completes high school, and about 45 percent enters college. It is estimated that over 1000 new books are published worldwide each day. Since World War II, the number of scientific journals and publications has doubled every ten to fifteen years. The only feasible way to keep up with current scientific developments is through abstract journals or computerized literature searches. Nobel Prize winner Emilio Segré recently declared, "On K-mesons alone, to wade through all the papers is an impossibility!"

The output of goods and services in the United States has been doubling about every fifteen years since World War II; in fact, some materials (such as plastics and synthetic fibers) have doubled every four or five years. The typical teenager of the 1980s is surrounded by twice as many material goods and conveniences as his parents grew up with. With only 5 percent of the world's population, this country uses 30 percent of the energy and an even higher percentage of the world's material resources. Our extravagant consumption of petroleum is particularly serious; just since 1959, the United States has used more petroleum than the entire world production up to that time.

The wastefulness of our life style is nowhere more apparent than in our love affair with the automobile. A car is no longer a luxury in the United States; it is considered a necessity, and an entire generation has grown up looking forward to owning a "personal car." We drive to the store for a single loaf of bread—or if we run out of beer. We drive to work when public transportation is often available. We drive our children to school by car or by school bus, and on a lazy Sunday the family drives down to the local shopping mall. We have drive-in banks and drive-in restaurants; in California, there is even a drive-in funeral parlor, where mourners may pay their final respects without leaving their cars!

These material possessions do not seem to have brought any degree of real happiness. A recent survey of high school seniors found that half admitted to more or less regular drinking of alcohol and 10 percent smoked at least one pack of cigarettes per day. Six percent of our adult population admit to using sleeping aids at least once a week. The United States produces over 300 million pounds a year of synthetic organic medicinal chemicals, including 136 million pounds of gastrointestinal medication, 60 million pounds of central depressants and stimulants, and 25 million pounds of antibiotics. *The New York Times* has reported that the most widely prescribed medication in the United States today is Valium (ironically from the Latin *valeo,* which means "I am well"). Prescribed for "anxiety," the 60 million prescriptions written in 1974 sold for half a billion dollars at local drug stores and returned some $250 million to its manufacturer. Another tranquilizer, Librium, accounted for some 20 million prescriptions that same year. There may well be as many as 3 million individuals in the United States who are dependent on such prescription drugs. According to the federal Drug Alert Warning Network (DAWN), in a one-year period between 1976 and 1977, 54,400 persons sought emergency room treatment for physical and psychiatric problems involving Valium (usually the result of combining it with alcohol or other drugs), and the Federal Drug Commission is finally taking steps to control this blatant abuse.

THE MYTH OF AFFLUENCE

During the last half of the twentieth century, we have become accustomed to measuring the affluence of nations in terms of *gross national product* (GNP), the market value, determined from national income and product accounts, of all goods and services produced in a calendar year. The GNP is often given in *constant dollars* referred to a base year, taking into consideration any change in the value of the dollar due to inflation. The GNP per capita is sometimes considered to be a direct measure of the quality of life today; thus, if the GNP per capita in the United States is twice that of another country, we would supposedly have twice as many material things and hence twice as much "joy of living" as the inhabitants of that country.

Oriented to a concept of infinite resources and unlimited growth since the founding of our country, it is not surprising that we easily adopted the GNP as a measure of the standard of living. Recently, however, it has been criticized for a number of reasons. All production is treated equally; $1 million for cigarettes is considered the same as $1 million for bread. No deduction is made for pollution or other adverse environmental effects or for the wearing out of machinery. Certain items are not included, such as the value of care for one's own home and family; there is also a question as to whether some items included in the GNP, such as military expenses, commuting costs, advertising, and many government services, represent any real gain in affluence. It is certainly not a measure of happiness or even of true productivity, and there is a growing feeling that it should be reduced since it is also a direct measure of the rate at which we are using up our natural resources.

Economist Kenneth Boulding has described (in *Collected Papers,* Vol 3) our way of life as a "cowboy" economy:

> The cowboy economy typifies the frontier attitudes of recklessness and exploitation. Consumption and production are considered good things and the success of the economy is measured by the amount of THROUGH-PUT derived in part from reservoirs of raw materials, processed by "factors of production," and passed on in part as output to the sink of pollution reservoirs. The Gross National Product (GNP) roughly measures this thoughput.

The philosophy of an expanding economy became something of a religion in the United States. For example, in 1952, Eugene Holman, then president of Standard Oil Company of New Jersey (now Exxon), published an article in the *Atlantic Monthly,* declaring:

> For many years, I believe, people have tended to think of natural resources as so many stacks of raw material piled up in a storehouse . . . Now I think we are beginning to discover that the idea of a storehouse—or, at least, a single-room storehouse—does not correspond with reality. Instead, the fact seems to be that the first storehouse in which man found himself was only one of a series. As he used up what was piled in that first room, he found he could fashion a key to open a door into a much larger room . . .. The room in which we stand at the middle of the twentieth century is so vast that its walls are beyond sight. Yet it is probably still quite near the beginning of the whole series of storehouses. It is not inconceivable that the entire globe—earth, ocean, and air—represents raw material for mankind to utilize with more and more ingenuity and skill.

Acclaimed a "true captain of industry," Holman saw his article appear in abbreviated form in *Reader's Digest* and his portrait featured by *Time, Fortune,* and *Harper's Magazine.*

In that same year, President Truman's Materials Policy Commission stated as a fundamental concept the following:

> First, we share the belief of the American people in the principle of growth. Granting that we cannot find any absolute reason for this belief, we admit that to our Western minds it seems preferable to any opposite, which to us implies stagnation and decay. Where there may be any unbreakable upper limits to the continuing growth of our economy we do not pretend to know, but it must be part of our task to examine such apparent limits as present themselves. . . .

An important feature of the growth philosophy has always been the implied idea that anyone who does something for personal gain is also a benefactor of society. In his 1776 classic, *The Wealth of Nations,* Adam Smith wrote that, in the case of a business man,

> . . . he intends only his own gain, and he is in this, as in many other cases, led by an invisible hand to promote an end which was no part of his intention. . . . By pursuing his own interest, he frequently promotes that of the society more effectually than when he really intends to promote it . . . .

This concept of an "invisible hand" was a strong factor in the government's laissez-faire attitude toward American business that lasted through the robber-baron days of the nineteenth century and the two world wars of the twentieth century. In fact, this attitude persisted up to a few years ago, as witness the often misquoted remark by Charles E. Wilson, a former president of General Motors, who during confirmation hearings as Secretary of Defense under President Eisenhower declared, ". . . what's good for the country is good for General Motors and vice versa."

Today this concept has been extended to international corporations operating on a global scale. For example, one of the partners in an international investment banking firm recently stated, ". . . working through great corporations that straddle the earth, men are able for the first time to utilize world resources with an efficiency dictated by the objective logic of profit."

Only in recent years has there been any serious question as to the desirability of further growth and whether, in fact, there are upper limits. In 1953, Samuel H. Ordway, Jr., in his *Resources and the American Dream,* presented a Theory of the Limit of Growth which assumed that basic resources could come into such short supply that rising costs would make their use in additional production unprofitable, and growth would then cease. Much more publicized was the 1968 formation of The Club of Rome, a group of thirty individuals from ten countries who met at the suggestion of Dr. Aurelio Peccei. These men initiated a computerized study, later published under the title *The Limits to Growth,* which concluded that, if present trends in world population, industrialization, pollution, food production, and resource depletion continue unchanged, the limits to growth on this planet will be reached within 100 years. They also concluded that the most probable result will be a rather sudden and uncontrollable decline in both population and industrial capacity.

The situation facing us today is described by Kenneth Boulding in his book, *Economics of Pollution:*

> When we contrast quantity with quality, we are suggesting that the bigger is not necessarily the better. For a baby, growth in weight is evidently desirable; for the adult, it simply means that he is turning into fat. For the poor, growth in income is entirely desirable; for the rich, it may simply mean corruption and luxury. . . . The enormous increase in the Gross National Product in recent years immensely exaggerates the increase in welfare (or affluence). If we were to deduct pollution, education, health, and commuting, as costs, from the Per Capita Disposable Income, the rise in the last forty years might look very modest indeed. It is very doubtful whether Americans are as much as twice as rich as their grandfathers, which is clearly a modest rate of real economic growth.

Today's energy crisis shows that the developed countries are, in fact, pushing their limits of growth. The "cowboy" economy is now beginning to give way to a "spaceship" economy, one where it is recognized that the reservoirs of energy and raw materials are finite, and that the reservoirs of pollution are also finite and cannot accept input too quickly.

A critical question throughout this book is to what degree the limited supplies and greatly increased costs of energy will affect growth in the United States. The cost of energy enters into the cost of everything we have. In addition, petroleum and natural gas are themselves chemical raw materials for the manufacture of plastics, fibers, detergents, drugs, and ammonia fertilizers. Any limitation of the amount of petroleum and natural gas available to us tends to limit the quantities of everything else we produce; it follows, of course, that any increase in the cost of petroleum and natural gas will increase the cost of everything else we produce.

Some idea of the effect of higher energy costs can be had by taking a look at our national per capita GNP in recent years, corrected for inflation. Between 1950 and 1973, the GNP increased about 4 percent a year in constant dollars, indicating a corresponding increase in our material affluence. In 1974 and 1975, it actually decreased; in 1976, it bounced back to the preembargo value, and there were modest increases during the next three years. Preliminary figures for 1980 indicate another drop. When the GNP is corrected for higher taxes, we find that there has been essentially no improvement in the standard of living in the United States since the 1973–74 oil embargo.

JET LAG

The rapid economic growth witnessed since World War II has resulted in an interesting phenomenon among a large percentage of our population. An entire generation under the age of 40 has never been exposed to anything but an affluent and continuously growing economy. It is as if they had been on a fast jet plane to places of ever-increasing material wealth. This is the only life they have known, and they simply regard increasing affluence as one of the inalienable rights granted by our society. Many Americans are suffering from a sort of *jet lag;* they cannot recognize that the 200-year trip is over. "It's all the fault of OPEC!" "It's really just the big oil companies!" "But you said we'd get a snowmobile this Christmas!"

Jet-lag attitudes make it very difficult to do anything really constructive about the serious problems facing the United States. Much of our population still believe there is no real energy shortage, that the problems have been created by OPEC and/or the oil companies as an excuse to raise prices. How can our Congress impose or enforce higher gasoline taxes, conservation measures, and possibly some kind of rationing for petroleum products when such a large fraction of their constituents believe these to be entirely unnecessary?

How long will it take for us to recognize that we can no longer support a throw-away economy? For years, automobile manufacturers have changed models every year at tremendous cost and with only one purpose—to convince customers that they should trade in their perfectly functional cars for the latest chrome-trimmed gas guzzler. Clothing styles for men, women, and even children are changed every year to make last year's garments noticeably out of fashion: skirts are raised, lowered, or even slit; lapels are broadened or narrowed; neckties change from narrow to wide and back to narrow after the older ones are discarded. An army of high-priced hustlers are employed to make the public feel dissatisfied with what they have, to make them rush out to buy the latest model.

An even more serious example of jet-lag thinking is the widely held concept that the United States remains a nation with huge untapped resources only waiting to be developed by advanced technology. Today we are importing 38 percent of our petroleum as well as very large quantities of critical raw materials such as iron ore, aluminum ore, lead, platinum, and chromium. Millions of our people are unemployed, and much of our industrial equipment is out of date. Although the "people-limited" culture of our first two centuries has become a "resource-limited" culture, we continue to accept millions of immigrants, both legal and illegal.

The situation has been made worse by a series of administrations and an entrenched bureaucracy in Washington who seem unable to face the facts. They have failed to realize (or perhaps been afraid to admit to themselves and to us) that decreasing resources spread over an increasing population can mean only one thing: less for everyone. The heyday of our material growth is over. We are not in a recession but rather at the beginning of a new way of life limited and controlled by our natural resources. It's a new ball game, one that demands a long hard look at pork-barrel politics and an end to our jet-lag thinking.

In the past, the United States has been very fortunate (or perhaps very unfortunate) in that we could tolerate an extremely wasteful economy and still have plenty left to send around the world in the form of foreign aid. We could literally eat our cake and have it. Today we are faced with difficult choices and often forced to seek out the best of several rather unsatisfactory solutions.

The question of future generation of electricity in the United States is a good example. Several options are open to us: (1) Continue to use OPEC oil regardless of price. (2) Convert to coal on a large scale, regardless of its serious air pollution and solid waste disposal problems. (3) Convert to nuclear power with its attendant hazards. (4) Reduce our demand for electricity. None of these is entirely satisfactory, but this is the type of choice we shall face increasingly in the future.

It is helpful to remember that other nations are confronted by different situations which may involve entirely different choices. France, for example, is basing almost her entire future generation of electricity on nuclear power and is very actively studying the fast-breeder reactor, a concept that the United States has placed on the back burner. The French government has been criticized for this emphasis on nuclear power in view of the potential hazards and the problems of radioactive waste. Yet, with practically no petroleum and only very limited supplies of coal, the choice for France was really quite simple: either develop nuclear power or else remain dependent on imported fuels for electric generation.

For the United States, on the other hand, it is becoming more and more obvious that the easiest and best choice is to conserve. While this answer is quite contrary to jet-lag thinking (and thus very unpalatable to our politicians), the truth is that our economy has plenty of fat to spare; a streamlined United States, with less anxiety about material gain, would continue to represent a standard of living unmatched in the history of the world.

THE TRAGEDY OF THE COMMONS

Probably the first and certainly one of the clearest descriptions of the basic problem facing the world today was presented by a professor of political economy, the Reverend William F. Lloyd, in a lecture delivered at the University of Oxford in Michaelmas term 1832. Using the analogy of the village common, where local farmers were allowed to graze their cattle, Lloyd asked:

> . . . Why are the cattle on a common so puny and stunted? Why is the common itself so bare-worn, and cropped so differently from the adjoining inclosures?. . . The difference depends on the difference of the way in which an increase of stock in the two cases affects the circumstances of the author of the increase. If a person puts more cattle into his own field, the amount of the subsistence which they consume is all deducted from that which was at the command of his original stock. . . , and he reaps no benefit from the additional cattle. . . . But if he puts more cattle on a common, the food which they consume forms a deduction which is shared between all the cattle. . . and only a small part is taken from his own cattle. . . .

In 1968, the concept of "the tragedy of the commons" was reintroduced by Garrett Hardin, with the word *tragedy* referring to the inevitable fate that befalls any form of common property. For example, take the case of pollution. Suppose a particular manufacturing operation produces an undesirable waste stream, which can be cleaned up only at the cost of a substantial amount of money. From the owner's point of view, the best thing to do is quietly dump the wastes into the nearest lake or river. This leads, of course, to degradation of the environment shared among all those living in the area, but the cost to the manufacturer is very small. There is no natural economic incentive to reduce pollution until it is required by law.

WHO'S IN CHARGE?

Closely related to jet-lag thinking is the widespread lack of confidence in our government today. There are many scarce commodities in the world but none so scarce as competent leadership. Washington's apparent inability to face hard facts and to differentiate between cause and effect has furthered a tendency to treat symptoms rather than disease and to rely on short-range "cosmetic" adjustments—anything to avoid the often unpleasant nitty-gritty decisions that are needed. Government officials have been acting as if they can no longer depend on the response of our citizens.

There is no question that our crisis of confidence began as a direct result of the Vietnam War, which was sold to the American public in 1964 on the basis of a probably fictitious Tonkin Gulf incident. Some of our young men fled the country to avoid service in Vietnam; others enrolled in college to obtain deferment and then "demonstrated" while their less fortunate comrades were being killed. Withdrawal from Vietnam was followed by Watergate, Koreagate, and Abscam. In the nationwide polls of 1980, one of the most popular presidential candidates turned out to be "none of the above," and an incredible number of eligible voters now remain home on Election Day.

Nowhere has the lack of competent leadership in the United States been more apparent than in our response to today's energy crisis. When the Arabs used their "oil weapon" during the Yom Kippur War, and placed an embargo on all oil shipments to the United States, there was confusion that bordered on panic. For years experts had been warning us that our oil reserves were limited and, in fact, domestic production of crude oil had begun to decrease after 1970. This did not restrain our appetite for oil, and imports rose sharply during the early 1970s. By the spring of 1973 (when Arab oil still represented only about 6 percent of our total demand), our total imports from the Middle East were *five times* what they had been in 1970. Small wonder that the Arabs saw this as an opportunity to try to alter our policy on Israel and lay the groundwork for a hefty increase in the price of their oil!

In retrospect, the embargo was actually the best thing that could have happened to the United States. It forced us to recognize that there was, in fact, a real energy crisis; it forced us to recognize and begin to do something about our wasteful energy consumption; it forced us to look for alternate domestic sources of energy, and—most important—it made us aware of the danger of relying on unstable regions of the world for our supplies of critical materials.

In addition to the energy crisis, the United States is facing problems of inflation, unemployment, immigration, foreign aid, and how to supply our needs without destroying the environment. The solutions will require the genuine cooperation of government, business, labor, and the public. How long will it be before we get our act together?

Chapter Two

The Population Explosion

It is generally agreed that the first *hominids* appeared in Africa some five million years ago. Unlike their predecessors, the apes, these creatures walked on two legs and never used their hands to bear their weight. Being reasonably intelligent, they found many new uses for their free hands, and eventually this led to toolmaking, the distinctive ability of mankind. (Note that free hands would have been of no value without the intelligence to use them; the kangaroo also has free hands, but these have simply been allowed to atrophy.)

The second important characteristic of the hominid was his increasing brain size. Although the first hominids had a brain only about 20 percent larger than the 500-cubic-centimeter capacity of the gorilla, the next two million years saw their cranial capacity gradually develop to 900 cubic centimeters for primitive man, *Homo erectus.* Another 2.9 million years brought this to the present average of about 1450 cubic centimeters for modern man, *Homo sapiens.*

Estimates of prehistoric world population are very approximate. Unlike the gorilla who literally ate his way through the forests, primitive man was a carnivore whose existence was primarily dependent on hunting, with limited gathering of seasonal fruits, nuts, berries, and roots. Prehistoric population figures are estimated from population densities of primitive cultures that have continued to live at this Paleolithic (old stone age) level into modern times, such as the aborigine population of Australia. Values given in Table 1 could easily be several times too high or too low but will serve as a general guide.

Figure 2. Vietnamese "boat people" in the South China Sea. (United Press International)

Table 1. Estimated Prehistoric World Population*

Years Ago	Cultural Stage	Estimated Population
1,000,000	Lower Paleolithic	125,000
300,000	Middle Paleolithic	1,000,000
25,000	Upper Paleolithic	3,340,000
10,000	Mesolithic	5,320,000
6,000	Neolithic	86,000,000
2,000	Christian Era	133,000,000

*Prof. E. S. Deevey, *Scientific American,* Sept. 1960, p. 195.

When the last ice age began, about 75,000 B.C., arctic conditions wiped out a large portion of man's comfortable habitat. Fortunately, *Homo sapiens* was able to learn to live and love in a cold climate. Fire had been known for many thousands of years and was used for cooking, warmth, and to discourage predatory animals. Protective clothing was made from the skins of animals. When the ice finally retreated, bands of hunters discovered a land bridge that existed in the Bering Strait, and within a few centuries mankind spread throughout the continents of North and South America. By this time, other groups had made their way to Australia by way of the Indonesian Archipelago, and the range of mankind had essentially doubled.

Early man existed in small nomadic tribes of at most a few dozen families. Life was very precarious. Dependent on hunting, they wandered across the land seeking areas where game and other forms of food were plentiful. At no time was there a substantial surplus of food beyond the needs of the hunter and his immediate family. Periodic famines were common, and for hundreds of years the population density showed little change.

From about 6000 to 4000 B.C., the Neolithic (new stone age) agricultural revolution occurred throughout the world as man learned to domesticate plants and animals. Wheat and barley became important crops in the Old World; in the Americas, the staples were corn and beans. Planting, weeding, and harvesting were largely "women's work," and girl babies were now welcomed. Cows and sheep provided milk, meat, and wool. With the occasional availability of surplus food, man moved out of caves and into more permanent camps and settlements. By about 3000 B.C., hunting was of relatively small importance to most of the world's population.

Even with the Neolithic revolution, population growth was at first very slow. Beginning about 4000 B.C., population doubled every 1000 years until about 1000 B.C, when a more rapid increase occurred during the height of the Greek and Roman civilizations. After the collapse of Rome, population actually decreased for a time and, from the Dark Ages through about 1700 A.D., was again doubling only every 1000 years. Both famine and disease played large parts in limiting growth during this period; for example, the Black Death (bubonic plague) of 1348 killed one quarter of all the inhabitants of Europe.

THE LAWS OF GROWTH

At this point in the story, it might be helpful to take a brief look at the basic laws of growth which affect population.

In *arithmetic growth,* a quantity always changes by a constant amount in a given period of time. Pure arithmetic growth follows the numbers 1, 2, 3, 4, 5, and so forth. One of the best examples is immigration quotas; for example, in recent years the United States has allowed close to 1 million legal immigrants each year, which has resulted in a corresponding yearly increase in our population. The relative importance of arithmetic growth becomes smaller as the population increases: one million added to ten million is quite significant; 1 million added to today's more than 200 million is not so visible.

Arithmetic growth can be positive, negative, or zero. A well-known example of negative population growth resulted from the Irish potato famine of the middle nineteenth century, when severe food shortages led large numbers to leave that country and its population decreased sharply for several years. A graph of arithmetic growth is a straight line on ordinary (arithmetic) graph paper.

In *geometric growth,* a quantity always changes by a constant percentage in a given period of time. Examples of geometric growth include world population growth and consumption of raw materials. The number of babies born each year is largely dependent on the number of women of child-bearing age; an increase in world population increases this number and (other things being equal) increases the number of babies. In other words, the number added in a given time increases as the population increases. Pure geometric growth follows the numbers 1, 2, 4, 8, 16, and so forth. Proof of geometric growth is indicated by a straight line on semilogarithmic graph paper (that is, graph paper that provides a logarithmic scale vertically and an arithmetic scale horizontally).

Geometric growth can also be positive, negative, or zero. World population growth in recent years is a good example of positive geometric growth. A number of people who are deeply concerned about this growth feel we should strive rapidly to reach "zero population growth," which translates roughly to slightly more than two children per couple. As we gradually consume our total resources, it is expected that a period of negative population growth will occur.

Obviously geometric growth under certain conditions may produce very rapid increases in numbers. This concept can be carried to ridiculous extremes, of course, such as estimating how long it will be before there is standing room only on the earth—or calculating that within a few years the descendants of a pair of fruit flies will weigh more than the total weight of the earth. There are always negative checks and balances which produce an upper limit to growth. This, in fact, is exactly the situation we see in the United States today, where our rapid postwar growth is leveling off as we push the limits of our raw materials (particularly our energy supplies).

The S-Curve of Growth

Mathematical growth laws are useful tools for explaining growth curves, but they must be used with considerable caution, particularly when trying to predict future growth. In the real world, nothing follows an exact growth curve and variations from year to year will depend on circumstances. For example, the population growth of the United States slowed significantly during the depression years of the 1930s. An unexpected "baby boom" in the 1950s was followed by a rapid decline in fertility beginning in the 1960s with the development of the "pill" and accompanied by women's lib and changing life styles that de-emphasize the traditional concepts of marriage and family.

A long-term look at the growth of a particular quantity usually shows that this can be divided into three general periods. First, there is a period (which may be quite long) where there is very little growth and the numbers remain almost constant. This is followed by a period of rapid geometric growth where the numbers increase rapidly with time, but in a third period negative checks cause the numbers to level off or even to decrease. When these numbers are plotted on semilogarithmic graph paper, a typical S-curve is revealed.

A typical S-shaped growth curve is illustrated in Figure 3, which gives the population of Switzerland for the past 800 years. Note the essentially station-

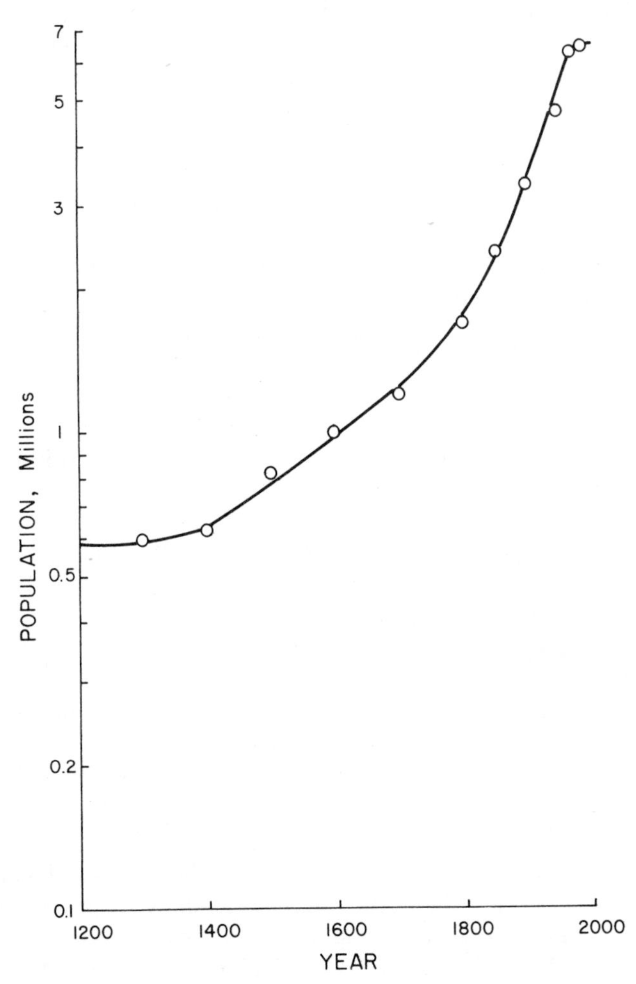

Figure 3. Population of Switzerland.

ary population from 1200 to 1400 A.D., followed by a modest growth until 1800 A.D., when the population started doubling about every century. In 1960, however, this growth essentially came to a halt, and the population has now leveled off to a constant value.

Doubling Time

Also important to our understanding of the growth laws is *doubling time,* the length of time required for a given number to double in size. This is usually calculated by the "rule of 70," which states that the doubling time of a number is approximately equal to 70 divided by the yearly percentage increase. A population increasing at a rate of 2 percent per year, for example, will double in approximately thirty-five years.

MODERN POPULATION GROWTH

A sharply increased rate of population growth throughout the world commenced shortly after the industrial revolution. During the eighteenth century, population began increasing at a rate of about 0.4 percent per year, corresponding to a doubling time of under 200 years. This continued through the nineteenth century and went up to 0.8 percent per year for the first half of the twentieth century—a doubling time of less than 100 years.

Following World War II, modern methods of disease control were introduced throughout the world. The use of DDT (dichlorodiphenyltrichloroethane) and other insecticides greatly reduced the incidence of malaria and other tropical diseases. On the island of Ceylon (now Sri Lanka), for example, its use actually reduced the death rate to half the previous number. Antibiotics such as penicillin became available; BCG vaccine (bacillus Calmet-Guérin) was introduced to prevent tuberculosis in many parts of the world; smallpox was entirely eliminated. Better drinking water, improved sanitation, and wider medical care further reduced deaths by disease. Meanwhile, the birth rate in many countries remained essentially unchanged, so that the population growth rate turned into a population explosion. Since 1950, the growth of the world's population has been just under 2 percent per year, which equals a doubling time of about thirty-seven years.

Put another way, mankind required several million years to build up to a total population of one billion (1000 million) at the start of the nineteenth century. In another 120 years, the population reached two billion. Adding the third billion took thirty-five years, the fourth billion only another fifteen years, and—at today's rate of growth—the fifth billion will take twelve years. Unless the growth rate decreases, the close of the twentieth century will find some 6.6 billion inhabitants on spaceship earth!

Without question, the chief problem facing the world today is the tremendous growth in population which has occurred since World War II. We now gain approximately 230,000 new people each *day*. Most of this increase is taking place in the undeveloped countries, where a large portion of the population is already inadequately housed and fed. Unless this growth can be stopped in the near future, which seems quite unlikely, mass starvation in many new areas will be just around the corner.

The Demographic Transition

To comprehend these figures, it is necessary to consider the birth rate, the death rate, and how these have changed through the years. The *crude birth rate* of a country is simply the total number of births occurring in a given year, divided by the population; it is normally expressed as births per thousand population. The *crude death rate* is, of course, the total number of deaths occurring in a given year per thousand population. The difference between the birth rate and the death rate is the actual population growth, or *crude rate of natural increase*. This does not include changes resulting from immigration or emigration. Table 2 provides some interesting contrasts over the past forty years among the twelve most populated nations of the world.

For a sexually active population that practices no form of birth control, the crude birth rate may be as high as fifty per thousand. Several hundred years ago, all countries had birth rates ranging from about thirty to fifty per thousand. This was necessary for survival, since correspondingly high death rates in those times resulted in only a modest rate of natural increase. The industrial revolution brought a decline in death rates in the nations of Northern and Western Europe, starting in the middle of the eighteenth century. For some 100 years birth rates remained high, resulting in a population growth unprecedented in human history. Beginning about 1850, however, the birth rates also began to decline until today the birth rates in these countries range from twelve to fifteen per thousand with death rates slightly lower.

This entire process has been called the *demographic transition*. It takes place in three stages: (1) a period of high fertility (birth rate) and high mortality (death rate); (2) a period of declining mortality with high or medium fertility and rapid population growth; and (3) a period of low fertility and low mortality with low or zero population growth.

In characterizing the nations of the world, it has become customary to divide them into two categories, the *developed* (that is, more industrially developed or "rich") nations and the *undeveloped* (underdeveloped, develop-

Table 2. Recent Growth of the World's Twelve Most Populated Nations*

	Crude Birth Rate (per 1000 population)			Crude Death Rate (per 1000 population)			Crude Yearly Natural Increase		
	1938	1950	1978	1938	1950	1978	1938	1950	1978
China, People's Republic	35–45	42–47	21–26	30–40	22–27	8	0.3–0.5%	1.5%	1.6%
India	45–46	40–43	33–34	31–33	27–30	15	1.2–1.5	1.2	1.9
Soviet Union	31	27	18	18	10	10	1.3	1.7	0.9
United States	17.6	23.5	15	10.6	9.6	9	0.8	1.7	0.8
Indonesia	40–50	40	34–36	30–35	20	15	1.0–1.5	2.1	2.0–2.1
Brazil	44–46	42–45	31–33	30–35	18–21	8–9	0.9–1.6	2.7	2.2–2.4
Japan	27	28	15	18	11	6	0.9	1.6	0.9
Bangladesh	40–50	50	44–48	30–35	25–30	17–20	0.5–2.0	2.3	2.5–3.0
Pakistan	40–50	50	44–45	30–35	25–30	16–17	0.5–2.0	2.4	2.8–2.9
Nigeria	50–60	56	49–51	40–50	25–35	17–20	1.0–2.0	2.0	2.9–3.4
Mexico	44	46	36	23	16	8	2.1	2.8	2.7
West Germany	20	16	9	12	10	12	0.8	1.1	−0.1

*Data obtained largely from U.S. Bureau of Census and U.N. Demographic Yearbook.

ing, or "poor") nations. In general, the developed nations are those that have undergone both the industrial revolution and the demographic transition. These include all of Europe, the United States, Canada, Japan, Israel, Turkey, Australia, and New Zealand. The undeveloped nations constitute the rest of the world. Of course, this classification is arbitrary and there are some gray areas. The separation is useful, however, because the problems faced by the developed nations are entirely different from those faced by the undeveloped nations.

The developed nations contain about one fourth of the world's population. Their growth rate has decreased continuously since World War II to about 0.7 percent a year today. Several nations, such as the United Kingdom, Switzerland, and both East and West Germany, have populations that are now stable or actually decreasing. It is interesting to note that, among the developed nations, neither religious or political preference have any effect on the number of children produced. Catholic Italy and Spain, Communist Romania and Yugoslavia, and Protestant Northern Europe all have very similar growth rates; the yearly growth rate of the United States is 0.8 percent as compared with 0.9 percent for the Soviet Union.

On the other hand, with three fourths of the world's population, the undeveloped nations are growing at a rate of 2.1 percent a year. This rate of growth has been essentially constant since 1950, and there is no indication that it will decrease significantly in the near future. India, Indonesia, Bangladesh, Brazil, Pakistan, and Nigeria all have growth rates ranging between 2 and 3 percent. Mexico has been growing at a rate of over 3 percent for the past twenty-five years, and is is now the eleventh most populated nation of the world.

A critical question facing the world at this point is whether the concept of demographic transition is applicable to these newly developing nations, and whether declining birth rates will follow the recent spectacular reduction in death rates and resulting explosive population growth of the undeveloped world. Examination of the statistics available gives little cause for optimism. While some of these developing nations, such as Singapore, have substantially reduced both their birth rate and their rate of natural increase, in most others —including all the most populated nations—both birth rate and natural increase are still very high.

Demographic transition in the case of the developed nations has always been closely associated with industrialization, urbanization, improved education, and higher living standards. Many people now believe that the undeveloped nations will not be able to make the demographic transition until their standards of living are very substantially improved. Unfortunately, this is a chicken-or-egg situation: practically all the additional resources, which have been provided for the undeveloped nations since World War II, have been quickly absorbed by their population growth, so that the standard of living of the typical inhabitant has improved very little. With the developed nations now facing shortages of energy and raw materials, it is hard to foresee any significant increase in the aid given to the undeveloped nations.

Even if the undeveloped nations are eventually able to make the demographic transition, there is the question of how much time will be needed. If today's developed nations required over 100 years for their transition, is it reasonable to expect the undeveloped nations to accomplish this in twenty or thirty years? In the developed nations, we are now well aware of the fact that children are very costly to raise and that almost all children will survive to become adults. In the undeveloped nations, it has been customary to have four or more children in the hope that two or more will survive to produce grandchildren. Patterns such as this are not easily changed, and generations may be required before the idea of a small family becomes culturally acceptable in many regions of the world.

Population Pyramids

A major factor affecting the nature and growth of a population at any given time is its *age structure,* that is, the relative numbers of children, adolescents, adults, and elderly within that population. Age structure is usually illustrated by means of a population pyramid, which shows the relative numbers within the age groups. The shape of the pyramid for a developing country is quite different from that of one which has undergone the demographic transition.

If a population is stable, both the total population and the number within each age group are constant with time. The birth rate is equal to the death rate. Each year a certain number of babies are born, and these simply pass up the pyramid as they grow, with each age group decreasing in number according to its own mortality. A stable population is characterized by both low mortality and a quite low fertility. It is an "old" population, with a small percentage of children and a relatively high percentage of elderly.

No population is completely stable because there are always fluctuations from year to year, but several nations in Northern Europe are close to this condition. Note the example of Switzerland (Figure 4), where, except for a small decline in fertility in recent years, the population pyramid is remarkably constant from childhood through middle age.

A growing nation, on the other hand, has a "young" population, where children constitute a large percentage of the total; its pyramid consists of a broad base with rapidly decreasing age groups going up from this. A young

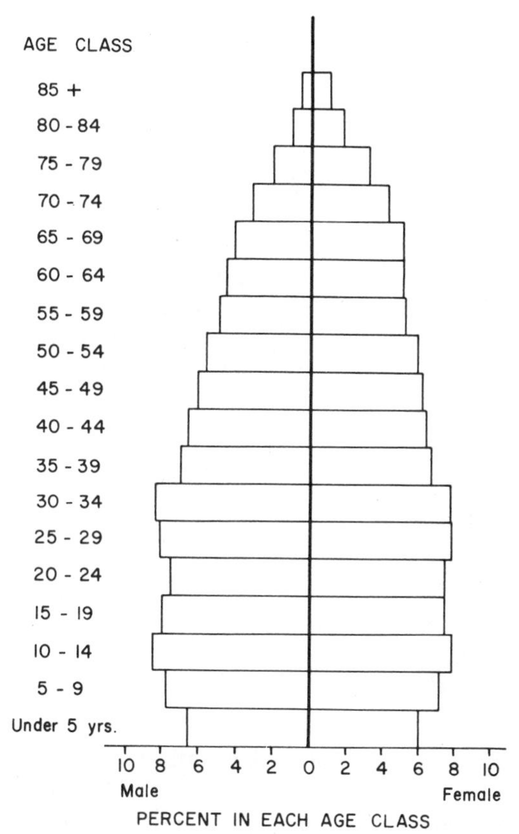

Figure 4. Population pyramid for Switzerland.

population is characterized by high fertility and low or medium mortality. Mexico (Figure 5) is a particularly good example of a growing nation, and you will note a large fraction of children with very small numbers of elderly.

The United States today (Figure 6) is an example of a maturing population. The decline in its birth rate during the depression of the 1930s resulted in a reduced number of adults presently in their 40s, but this was followed by the postwar baby boom which produced the large number of today's young adults. Since the 1950s, the birth rate has been steadily decreasing, but there may well be a modest increase as the postwar babies enter their child-bearing years.

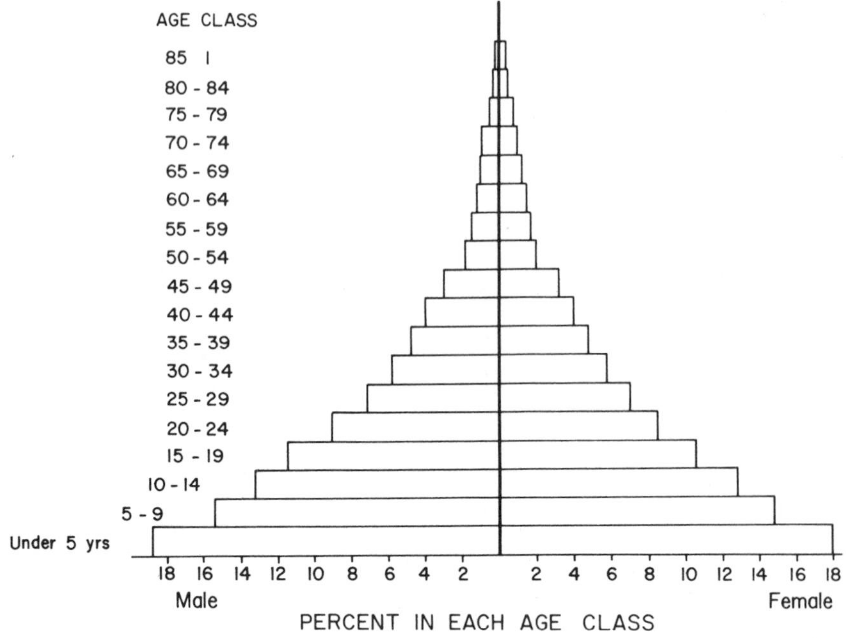

Figure 5. Population pyramid for Mexico.

Dependency Ratio

Closely related to the age structure of a nation's population is the *dependency ratio*—the ratio of the population under 15 and over 65 to the population aged 15 through 64. This is a rough measure of the number of persons too young or too old to work who must be supported by the current potential working population, assumed to be those between ages 15 and 64.

Typical values for both developed and undeveloped nations are given in Table 3. Note that in the developed countries less than one quarter of the population is under the age of 15, and there is a large fraction of elderly.* In these countries there are almost two workers for each dependent child or adult. In the undeveloped countries, with over 40 percent of the population under the age of 15 and a very small percentage of elderly, the dependency rate is much higher and there is slightly over one worker for each dependent individual.

Because of its postwar baby boom, the United States has a somewhat higher

*Japan is an exception because of the large number of persons in today's over-65 bracket who were killed in World War II.

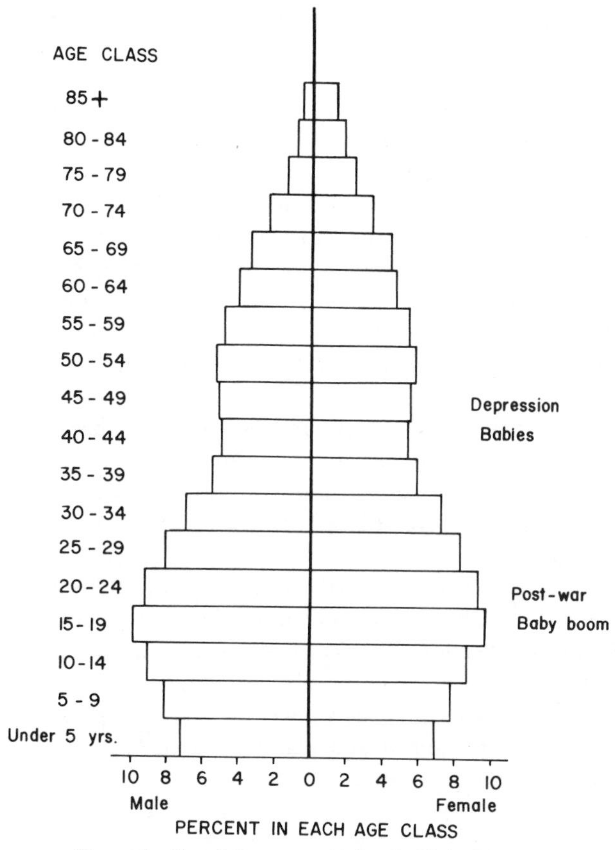

Figure 6. Population pyramid for the United States.

percentage of children and a somewhat lower percentage of elderly than the nations of Northern Europe. In the years to come, the percentage of children under 15 in the United States will gradually decrease and level off at a value of about 20 percent of the population. This will reduce the need for schools, pediatricians, toys, large houses, disposable diapers, and other child-oriented goods and services. The percentage of elderly, however, will be gradually increasing to 15 percent or more of the population. This will require a substantial increase in government services such as Social Security, housing, and medical assistance, as well as private nursing homes and other geriatric requirements.

The situation in the undeveloped world today is almost impossible to solve without strict limitation of population growth. Whenever over 40 percent of a population is under the age of 15, and when this fraction of the population is continuously increasing in size, a major part of the limited national income must be devoted to the struggle to provide minimum food, housing, and education for these dependents. Even with a modest expansion of social services, the average child will be no better off because the resources are spread over a continually growing number. As one cynic paraphrased a popular song in the 1930s, "the rich get richer, and the poor get children."

The problem of education in the undeveloped world is particularly serious because today's children are tomorrow's workers. Because of inadequate schools, most children in these countries are simply raised and educated in the spare time of the mother, whose main efforts must be directed to the work of a farm, small shop, or local industry. Some idea as to the magnitude of the problem can be realized from United Nations estimates that roughly one third of the adults in the world are illiterate; also, because of cultural traditions, a large fraction of the illiterates are female. In spite of the substantial quantities of money going into education, the number of illiterates worldwide has actually increased in recent years because of the exploding world population.

Another urgent problem faced by the undeveloped world is unemployment. A continuously increasing population means increasing numbers of people entering the job market for the very small number of available jobs. Even university graduates have severe problems finding employment in the undeveloped countries. This extremely high rate of unemployment produces much social misery and unrest and places an even heavier burden on the working fraction of the population.

Table 3. Dependency Ratio of Selected Countries

	Percent Persons under Age 15	Percent Persons over Age 65	Dependency Ratio per 100, Age 15–64
Japan	24	8	48
Switzerland	22	13	53
United States	24	11	53
West Germany	21	15	55
Sweden	21	15	56
Austria	23	15	61
Brazil	41	3	79
India	41	3	79
Bangladesh	44	3	87
Indonesia	44	3	87
Mexico	46	4	99

A direct result of high unemployment in undeveloped countries has been increasing urbanization. With few jobs available in rural areas, migration to the cities has produced urban slums which are doubling in size every five to seven years. These migrants remain "villagers" in an urban setting and cannot be readily absorbed into the urban social and economic system. Many of the larger cities of the world have shown fantastic growth; for example, it is estimated that the population of Calcutta, which reached 7 million in 1971, will be between 40 and 50 million by the year 2000.

Population Momentum

A very discouraging aspect of the problems faced by the undeveloped world is that population growth automatically begets more population growth. When a particular country has a large fraction of its population under the age of 15, even if the birth rate can be reduced to the "replacement rate" of slightly over two children per family, its population will continue to increase substantially as the large number of children grow into their child-bearing years. It has been estimated that, if the average number of children per family throughout the world could be reduced to 2.1 by the year 2000 (a very unlikely prospect), world population would eventually stabilize at about 8 billion—almost twice the present population.

Asia

Five undeveloped nations of Asia today represent almost half the world's population: China, India, Indonesia, Bangladesh, and Pakistan. Growing at a rate of 2 percent or more each year since 1950, they presently contribute an additional 40 million people to the world population per year. With no real indication of any reduction in this rapid growth rate, it is hard to see anything but disaster ahead for this region of the world.

The only one of these nations for which there can be present optimism is China. Even with a billion inhabitants, China has tremendous resources—most of them undeveloped. Effectively cut off from the West from the end of World War II until just recently, Maoist China had to raise itself by its bootstraps. Under a totalitarian government, food production was increased, the birth rate was decreased, and the standard of living was raised to the point where mass starvation has been eliminated. Birth control has become a national policy, with bonuses paid to families who stop at one or two children; the salaries of those who have a third child are reduced 10 percent until the child reaches the age of 14, and the third child is denied free education and medical care. Digestible "paper pills" are marketed in perforated sheets like postage stamps; special "vacation" pills and even morning-after pills are available. The immediate goal is to reduce population growth to 1 percent by 1981, with a further reduction to the replacement level by the end of the 1980s.

Nowhere are the problems of establishing birth control better illustrated than in the case of India. In the early 1970s, Prime Minister Indira Gandhi instituted a strong program of birth control based primarily on forced sterilization, although the authorities preferred to call it *family planning.* It was largely a failure: people actually left their villages and hid in the fields; wild rumors circulated throughout the nation. In October of 1976, there was a bloody riot in the town of Muzaffarnagar; some 50 to 150 people were killed when police opened fire on villagers protesting government attempts to recruit "volunteers" for sterilization. This was a major factor in the defeat of Gandhi's Congress Party in the national elections the following year.

Indonesia has limited oil supplies as well as the many problems of a rapidly growing population spread throughout its islands. India, Bangladesh, and Pakistan have almost 20 percent of the world population and very little in the way of natural resources. India has been able to slightly reduce its population growth but it remains much too high; Bangladesh and Pakistan both have growth rates close to 3 percent. With higher costs of energy and fertilizers now placing additional burdens on food production, it appears likely that starvation may be the eventual means of population control for these nations.

Africa

Africa has about one tenth of the world's population, and its rate of growth is the highest of any continent. Many of its nations, such as Nigeria, Ethiopia, Morocco, Zaire, Tanzania, and Kenya, have birth rates close to fifty per thousand. Even though the death rate of eighteen to twenty per thousand is twice that of the developed nations, Africa's present yearly growth rate is close to 3 percent.

Some of the nations of Africa are rich in natural resources while others have very little. The one natural resource that will become increasingly important here is hydroelectricity, since this continent has a large share of the world's undeveloped hydropower. While much of the hydro potential is poorly located, it must be considered an important resource for future development.

Agriculture throughout much of Africa is dependent on seasonal rains which have been unreliable in recent years. Over-grazing has also become a problem in many areas. It is hard to see anything but continued severe problems for much of Africa.

Latin America

With 8 percent of the world's total population, Latin America has a fairly high birth rate and a low death rate. Some nations, such as Argentina and Chile, have been able to reduce their population growth; others, such as Brazil, Venezuela, Mexico, and Peru, have growth rates close to 3 percent a year—among the highest in the world today.

From our standpoint, probably the most important Latin nation is Mexico with whom we share a 1400-mile border. The birth rate in Mexico is about thirty-six per thousand and, because of the large fraction of children, the death rate is only eight per thousand, one of the lowest in the world. The growth rate has increased from 2.7 percent per year in 1950 to the very high 3.3 percent in recent years. Mexico is now the eleventh most populated nation of the world, and its yearly population gain is now greater than that of the United States.

It has been estimated that the unemployment rate in Mexico is 30 percent, corresponding to some nine million jobless. In addition, some 600,000 or 700,000 new people enter the job market each year. Because of this population pressure, an estimated two million Mexicans each year illegally cross the relatively unguarded border to seek employment in the United States, with one million successfully avoiding capture—a number equivalent to over half the yearly population gain of the United States.

The recent discovery of large deposits of petroleum in Mexico may change this situation, and the increased resources for foreign trade and continued industrialization of this neighboring country may eventually help to solve a very serious population problem.

The United States

It has been estimated that the population of North America above the Rio Grande was 1 million in 1650 and 1.3 million in 1750. Since the North American Indian tribes were dependent on hunting, with a minimum of agriculture, it is probable that the population was fairly constant before 1650, and we can assume that the population of the territory that later became the United States was a million or slightly under when the white man arrived.

If we take a look at Figure 7, we see that starting with the first census in 1790 the population increased at a constant rate of 3 percent per year for about the next seventy-five years, corresponding to a doubling of population every twenty-three years. This rapid growth was obviously a combination of immigration, natural increase, and acquisition of new territory. The growth rate gradually slowed down after the Civil War and today is about 0.8 percent per

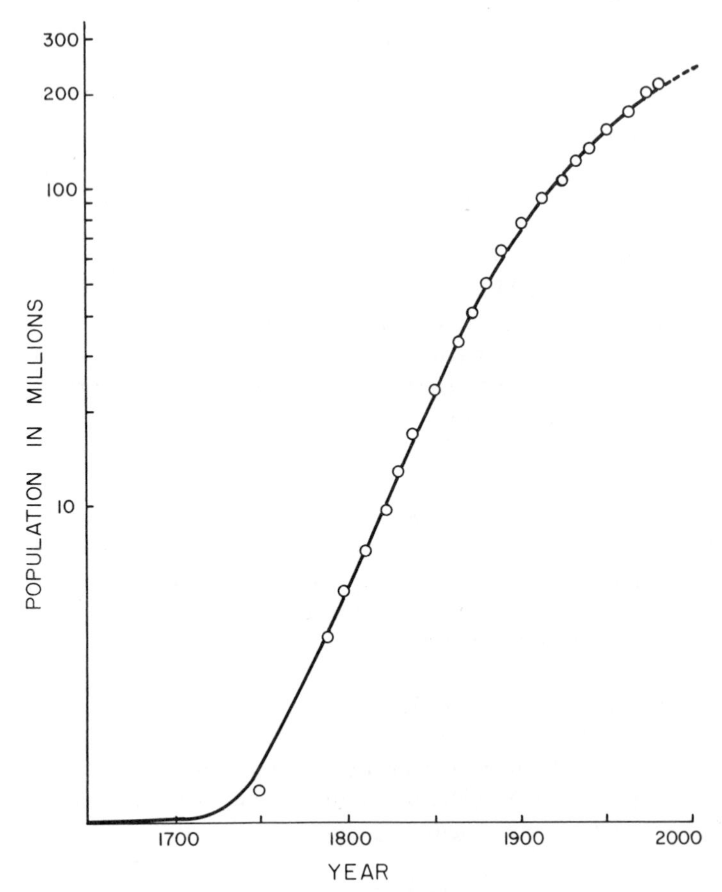

Figure 7. Population of the United States.

year, so that for the past century we have been on the leveling-off part of a typical S-shaped growth curve.

Future growth in the United States will depend on whether or not we restrict immigration and whether the present decline in fertility continues. If replacement reproduction can be achieved in 1985, for example, our population will continue to grow and will level off at about 300 million by the middle of the next century. It now appears that the 1980s may witness a modest increase in the birth rate with a corresponding increase in population growth. Our postwar babies are now having babies themselves—and marriage and family seem to be coming back into favor with many of today's young people. In addition, the growing Hispanic component of our population is increasing the overall birth rate, particularly in areas such as Florida and Southern California.

Immigration

As far as the United States is concerned, the most critical population problem is the question of immigration, both legal and illegal. Our situation is unique, since in the entire rest of the world immigration is very strictly controlled. Some countries, such as Switzerland, permit almost no immigration at all. Others, such as Australia, admit many immigrants but with restrictions: only certain ethnic groups are eligible, and they are carefully distributed throughout the country to avoid building up any local concentrations of minority groups.

As we have already noted, the problem of immigration is simply another example of jet lag. Most people in the United States have very mixed feelings about immigration. The Statue of Liberty stands in New York harbor and invites the hungry masses of the world to enter. Many of us are immigrants ourselves, or the children or grandchildren of immigrants. We have the most generous laws in the world: 290,000 legal immigrants are permitted each year plus exceptions for relatives, hardship cases, and "favored" groups. Why should there be a problem?

The plain truth of the matter is that the United States really has no immigration policy at all. Every time there is a refugee problem any place in the world, our Congress grants exceptions to the immigration laws. "Boat people" leave Vietnam in a steady stream, and many of them eventually make it to our shores. Haitians arrive in South Florida claiming to be political refugees and immediately come under the protection of our courts. When Fidel Castro permitted over 110,000 Cubans to leave the island, President Carter said they would be welcomed "with open arms"; violent demonstrations erupted in the refugee camps set up for them only a few days after their arrival.

To further illustrate how our immigration laws actually work in practice, consider immigration to the United States during fiscal 1978, as published in the Annual Report of the Immigration and Naturalization Service (INS). Listed are a total of some 600,000 legal immigrants, of whom only half were subject to numerical limitations. The others included parents, spouses, and children of United States citizens, 100,000 refugees from Indochina, 28,000 from Cuba, plus miscellaneous other categories of immigrants exempt from numerical limitations. Also reported were over a million returning resident aliens who were officially classified as "nonimmigrants" plus several million temporary visitors, many of whom probably have disappeared into the landscape.

In March, 1980, Congress passed the Refugee Act of 1980 to ". . . provide a permanent and systematic procedure for the admission to this country of refugees of special humanitarian concern to the United States, and to provide comprehensive and uniform provisions for the effective resettlement and absorption of those refugees who are admitted." (Of course, this also provided for further expansion of the Washington bureaucracy funded by our taxpayers by creating the Office of Refugee Resettlement.) While, theoretically, the Refugee Act limited the number to 50,000 maximum in any fiscal year, the President can unilaterally increase this number merely by stating that a special need exists. Less than three months after this Act was passed, there appeared to be agreement on all sides that it was unsatisfactory!

When the Iranian hostage crisis began in late 1979, our government wanted to know quickly how many Iranian students were present in the United States. The INS could not say! Before they were halted by a federal court, however, they were able to locate some 56,000 Iranian students, of whom 18 percent were in violation of their visa status. What about students from other countries? What about immigrants in general? No one really knows the exact figures, least of all the Immigration and Naturalization Service.

This did not come as a surprise to anyone who has had experience teaching foreign students. For years it has been perfectly obvious that a large number of the foreign students coming to the United States have no intention of ever returning to their native country. Some of them forge documents to gain entry; once admitted, they will do almost anything to avoid returning home. University graduate schools are filled with foreign students who have become perpetual graduate students, moving from one university to another to maintain their student visas. Many of them marry American citizens. Others seek an employer who will help them get the coveted "green card" (permanent visa status). Still others simply disappear into the population, and it is now estimated that about 10 percent of the nation's illegal aliens are visa abusers.

All of our other immigration problems are minor, however, in comparison with the question of Mexican immigration. While our legal yearly immigration quota from Mexico is 20,000 (and many feel this should be raised to 50,000), almost a million illegal Mexican immigrants are apprehended each year and sent back across the border. It is estimated that another million illegal Mexican immigrants are successful each year, but no one knows the exact number. In 1978, the Bureau of the Census estimated there were 7.4 million Hispanics in the United States illegally, most of them from Mexico. Former Immigration Commissioner Leonard Chapman estimated that one million United States jobs were held by illegals. As former Attorney General William B. Saxbe has stated, "they mock our system of legal immigration."

Legal immigration accounts for about 30 percent of our population growth each year. Even using a conservative figure for illegal immigration, there appears to be no question that *most of the growth in the population of the United States is coming from legal and illegal immigration.* By the end of this century,

Hispanics will almost certainly overtake blacks to become the largest minority group in the land. In addition to the increase by immigration, the birth rate of Hispanics in this country is more than twice that of whites and 60 percent higher than that of blacks. In terms of population, the United States is now the fifth largest Hispanic country in the world—coming only after Mexico, Spain, Argentina, and Colombia.

The situation of the island of Puerto Rico is particularly interesting because of its close ties with the United States. The birth rate of twenty-three per thousand and the death rate of six per thousand are both low for Latin America. While the immigration rate of this island of slightly more than three million persons is estimated by the U.S. Bureau of Census to be fifteen per thousand, Puerto Rico shows an overall growth rate of 3.2 percent. What is the reason for this?

The answer seems to be that Puerto Rico has considerable unemployment and—because of unrestricted migration—many of its young adults come to the mainland for work. Many of them marry and have children in the United States, returning home eventually after they have saved enough from their earnings. Since a large portion of the population is away from the island during child-bearing years, its birth rate is reduced but then reappears later as an immigration rate.

Continued growth of our Hispanic population, whether legal or illegal, poses many serious questions for the United States. What will be the long-term impact of this massive Hispanic inflow? What will be the eventual status of Puerto Rico? Are we headed toward constitutional recognition of two languages and two cultures within this country? In view of the many problems of bilingual nations such as Belgium, Yugoslavia, and French Canada, we should be extremely careful not to get into a similar situation.

Serious minority problems are already developing in our southern states. Riots by blacks in Miami in the spring of 1980 were at least partially caused by their feelings that—although they had been residents of the United States for several centuries—they were being ignored and that most of the available resources were being devoted to newly arrived immigrants from Cuba and Haiti. Public schools in Texas have refused to accept the children of illegal immigrants. The 1974 Supreme Court decision on bilingual education is being widely attacked; many view this as deliberate encouragement for a separate Hispanic culture within the United States.

Of course, the problem of illegal immigration is further compounded by employers who take advantage of the situation. At present there are no penalties for employing illegal immigrants. That they are "docile" employees with no union affiliation and minimum benefits encourages unscrupulous employers to pay less than minimum wages and even require extra "volunteer" work hours at no pay. The illegals are perfectly aware that, if there are any questions or complaints, one word to the immigration authorities will start them on the road back home.

Is there a solution to the problem of illegal immigration? The answer is very simple: remove the motivation for illegal immigration by forbidding the employment of illegal aliens. Practically every country in the world has strict rules against the employment of noncitizens; for example in West Germany, an employer can be fined up to $20,000 for employment of an illegal alien. Other countries have similar laws. Washington finally seems to be facing the fact that something has to be done, possibly through a national identity card or through a computer system tied into the Social Security system. Although civil libertarians see this as just one more invasion of personal privacy by government, with several million United States citizens unemployed at present, it seems only a question of time before action is taken to restrict the employment of noncitizens.

Optimal Levels of Population

It is very difficult for a country to establish a desirable or *optimal* population level since each of us has different needs and expectations. As a minimum, it might be argued that a country should at least have the resources to provide adequate food, housing, education, and medical care for each of its inhabitants; when and if this point is reached, it will be impossible to set an arbitrary limit as to how much more is optimal. For example, the rapid population growth of the United States during the nineteenth century brought us many material "improvements" such as railroads, electricity, mass-produced finished goods, and the automobile. It also brought us noise, pollution, and crowded cities. Today, many people feel our materialistic standard of living has not been worth the environmental price we have had to pay; they consider the United States to be already grossly overpopulated. Others are still strongly growth-oriented and seriously believe that only through continued growth can we continue to provide more things for more people!

Even if we define optimal population as being that level beyond which further increase would no longer improve the quality of human life, we are still faced with the problem of deciding exactly when a nebulous quality of life begins to decline. There is also the question of density. Most of us would probably agree that many areas of the United States, such as New York City and Los Angeles, are now too crowded for enjoyable and healthful living. There are, however, tremendous areas of the Midwest and Rocky Mountain states that could comfortably accommodate a much higher population.

Today we import over one third of our petroleum, and one person might well argue that from an energy standpoint this indicates we already have more people than is desirable, assuming continuation of our wasteful energy habits. Another could respond that all we have to do is reduce our per capita consumption to the level of Western Europe—and there would be no problem. From the standpoint of food supplies, we could easily support twice our present population by making a few healthful adjustments in our eating habits. (Again, many would reply that they like the way they eat and do not want to give up any of their steak or hamburger.)

For a nation that has undergone the demographic transition, then, perhaps the best definition of optimal population might be *that population at which concerted efforts are made to restrict further population growth.* Presumably this represents the point at which a majority of the people feel that further growth would diminish the quality of their lives and the lives of their children. By this definition, several nations of Europe have already reached their optimal population. The United States has not as yet reached that point, but as we continue to spread diminishing resources over an expanding population, the day is sure to come when there will be a concerted demand for an end to our population growth.

THE CONTROL OF POPULATION

Throughout the ages, disease, famine, and war have all played a part in the control of population growth. In times of famine, infanticide, deliberate killing of the aged, and human sacrifice were not uncommon among primitive societies. Birth control and abortion by one means or another have been practiced almost universally and have been developed to such a degree that the problem of population control has now been solved from at least a scientific point of view.

From a cultural and religious viewpoint, the problem is far from being solved. Modern methods of birth control have been made available to much of the undeveloped world in recent years, in most cases without finding any wide degree of acceptance. In fact, the words *birth control* still arouse strong feelings on the part of many persons throughout the world. The well-known attitude of the Roman Catholic Church as well as the opinions of such divergent authorities as Thomas Robert Malthus and Karl Marx are widely quoted today. Cultural opposition to birth control remains very strong in some areas, even to the point of considering it a form of genocide in disguise. The entire picture is extremely complex, and there is still the real question as to whether birth control will be successful even in those nations that need it most.

To understand the situation, let us take a look at the origin of some of the basic ideas of population control.

Who was Malthus?

Essay on the Principle of Population as it Affects the Future of Society, which has aroused so much controversy through the years, was published in 1798 by Thomas Robert Malthus. Born in Surrey, England, in 1776, the son of a gentleman who prided himself on his advanced ideas, Malthus received most of his early education from private tutors. In 1784, he was sent to Cambridge University, where he had a distinguished undergraduate career and graduated in 1788 with an honors degree in mathematics. That same year he took holy orders and accepted a post as minister to a small parish in Surrey. In 1805, he became the first professor of political economy (and also modern history) in the English-speaking world at the East India Co. College in Hertfordshire, a position that he held until his death in 1834. He was happily married in 1804 and had three children, two of whom survived to adulthood.

Malthus' famous essay grew out of some discussions which he had with his father on the "perfectibility" of society. Many liberal writers of the day attributed the evils of society exclusively to such things as bad laws, corrupt governments, and greedy employers; thus, if these were reformed or abolished, they believed that suffering would disappear and man's very nature would undergo a radical organic change so that he would no longer be his own worst enemy. Malthus argued that the realization of a happy society would always be hindered by the miseries resulting from the tendency of population to increase faster than the means of subsistence. Impressed by his arguments, Malthus' father suggested that he put his ideas in writing—and then recommended publication of the finished manuscript.

Malthus' principle of population was based on the following two assumptions:

> First, That food is necessary to the existence of man.
>
> Secondly, That the passion between the sexes is necessary and will remain nearly in its present state.
>
> The power of population is indefinitely greater than the power in the earth to produce subsistence for man. Population when unchecked increases in a geometrical ratio 1, 2, 4, 8, 16, 32, 64, ... Subsistence increases only in an arithmetical ratio 1, 2, 3, 4, 5, 6, 7, ... Thus while population increases to 64 times its original size, subsistence will increase to only 7 times its original size. Thus the increase of the human species can only be kept commensurate to the increase of the means of subsistence by the constant operation of the strong law of necessity acting as a check upon the greater power. This implies a strong and constantly operating check on population from the difficulty of subsistence. This difficulty must fall

> some where and must necessarily be severely felt by a large portion of mankind.

Why did this small work by Malthus cause such a furor? His arguments were not original, having been expressed previously by many authors. The assumption that population increases geometrically was based on a consideration of the United States, which was then undergoing very rapid growth; however, growth rates in Europe were much smaller, and in some nations such as France the growth rate was already decreasing. The assumption that subsistence increases arithmetically ignores any contribution from science; the nineteenth century would witness substantial improvements in agriculture, and the use of selected strains of crops such as hybrid corn in the twentieth century would double and triple the yield from an acre of land.

To understand the reception given to Malthus' *Essay,* we must remember that the end of the eighteenth century was a time of economic hardship and social despair for European nations. In 1789 had come the French Revolution with the fall of the Bastille, the execution of the king and queen, the "Declaration of the Rights of Man," and the Reign of Terror. Napoleonic wars ravaged half of Europe, and there were rumors that France planned an invasion of England. Movements for social and political reform received much sympathy from the intellectuals, and there was much speculation about the possible improvement of society.

In England, British suspicion of French institutions and French intentions boiled in widespread alarm and hostility. The rulers of Britain saw "republican" feelings sweeping through the minds of the English working class. Those in the "Establishment" who feared radical social reform fought back hard against those who worked for it. The Habeas Corpus Act was suspended from 1794 to 1801, and there were many trials for high treason, often with savage sentences. But physical repression was not enough for those who feared reform; some means had to be developed to combat the new notions of "perfectibility" of man and society which were infecting intellectuals and workers alike.

Under these conditions, Malthus' *Essay* was a gift from heaven. Although it was probably not his intention, the *Essay* by Malthus was taken to be a political rebuttal against the French Revolution and the contemporary ideas of reform in England. Not only was society not "perfectible"; it was useless to undertake any major reform when humanity was always breeding up to the starving margin. The industrial revolution was just beginning to be felt in England, and many in the ruling classes felt that the views of Malthus relieved them of any responsibility for the deplorable conditions of the working classes; the latter had only themselves to blame, and it was thus not the negligence of their "betters" or the working policies of the country that were at fault. Proposals of social improvement were rejected on the grounds that an increase of comfort would only lead to an increase in numbers, so the last state of things would be worse than the first.

On the other hand, immediately following publication of his *Essay,* Malthus found himself attacked by humanitarians and social reformers all over England. This man who "defended smallpox, slavery, and child murder" was surely the most abused man of the age: "This vile and infamous doctrine, this repulsive blasphemy against man and nature"; "That black and terrible demon that is always ready to stifle the hopes of humanity"; "Unless Mr. Malthus can contrive to starve someone, he thinks he does nothing." To add insult to injury, Malthus had the impudence to marry and have three children after preaching against the evils of a family!

Even today the chorus continues, as witness the 1953 comment by a social scholar, "Malthusianism is a warning against all attempts to ameliorate the condition of society." In his 1954 book entitled *Marx and Engels on Malthus,* Ronald L. Meek stated:

> The theories of Malthus, now as always, are serving as weapons in the hands of people who, whether they are aware of it or not, are hindering the progress of mankind towards a fuller and more abundant life. . . . These doctrines have to-day become an important part of the ideological stock-in-trade of imperialism in its present state of crisis. . . .

Interestingly enough, the strongest attacks on Malthus have come from two extremes, the communists of the far left and the Roman Catholics of the far right. The official doctrine of the communist party is that there can be no population problem under socialism since each new individual brings another pair of hands to join the workers. Thus Karl Marx insisted that only faulty social organization (i.e., capitalism) causes poverty, and thus only good social organization (i.e., communism) can cure it. Marx's collaborator Friedrich Engels, the German socialist, also attacked Malthus while discussing the "myth" of overpopulation, as stated by Meek:

> The productivity of the land can be infinitely increased by the application of capital, labor and science. . . . Labor power grows together with population; . . .this immeasurable productivity, administered consciously and in the interests of all, would soon reduce to a minimum the labor falling to the lot of mankind. . . .

The attitude of the Roman Catholic Church is, of course, well-known. In a 1968 papal encyclical entitled *Humanae Vitae,* Pope Paul VI admitted that there is widespread fear that world population is growing more rapidly than the available resources, with growing distress to many families and developing countries. However, control of population by any form of "birth control" is still contrary to the teachings of the church.

> Each and every marriage act must remain open to the transmission of life. . . . The direct interruption of the generative process already begun, and above all, directly willed and procured abortion, even if for therapeutic reasons, are to be absolutely excluded as licit means of regulating birth. . . . Similarly excluded is every action which, either in anticipation of the conjugal act or in its accomplishment, or in the development of its natural consequences, proposes, whether as an end or as a means, to render procreation impossible.

Positive Checks

Malthus described two types of checks to population growth—"positive" and "preventive" checks. The positive or death-producing checks were those acting as a result of unrestrained population pressure, or in Malthus' words:

> Through the animal and vegetable kingdoms, nature has scattered the seeds of life abroad with the most profuse and liberal hand. . . . The germs of existence contained in this spot of earth, with ample food and ample room to expand in, would fill millions of worlds in the course of a few thousand years. Necessity, that imperious and all pervading law of nature, restrains them within the prescribed bounds. . . . Among plants and animals its effects are waste of seed, sickness and premature death. Among mankind, misery and vice. . . .

In primitive societies, the positive checks included disease, epidemics, famine, war, human sacrifice, feuds, witchcraft, and the deliberate killing or desertion of the aged. One of the most common forms of population control was infanticide. This had the advantage of being simple, positive, and it also permitted sexual selection: a hunting tribe could kill predominantly females, whereas an agricultural tribe might kill predominantly males. In most uncivilized groups, life was not held sacred until a child had gone through a formal ceremony of tribal recognition, often involving marriage. Until such time an individual was not considered a member of society, hence not even a human being. It could be killed without compunction, and the parents had full authority.

Although infanticide was openly practiced by primitive tribes, it was also quite common in Western Europe. Malthus referred to what he termed the "bad nursing of children," which was common in the eighteenth and nineteenth centuries, and which was really a disguised form of infanticide. In the cities at that time, it was common practice to leave babies with nurses or caretakers, sometimes referred to as "angelmakers," who gave the children gin or drugs to keep them quiet. The situation was so bad that Benjamin Disraeli in 1845 wrote, "Infanticide is practiced as extensively and as legally in England as it is on the banks of the Ganges; a circumstance which apparently has not yet engaged the attention of the Society for the Propagation of the Gospel in Foreign Parts." The lower classes left their newborn infants at foundling hospitals or orphanages which existed in all large cities. Since these institutions could accommodate only a small fraction of the infants they received, care was minimum and most infants were simply shipped out to peasant nurses in the country. Most of these children died within a short time, either from malnutrition or neglect.

In certain of the Italian hospitals, the mortality of the "foundlings" under one year of age was between 80 and 90 percent. In the early nineteenth century in Paris, over a third of all the children born were left to their fate as foundlings. In 1811, Napoleon decreed that foundling hospitals should be provided with a turntable, so that babies could be left without the parent being recognized or subjected to embarrassing questions. It was even seriously suggested that unwanted babies be painlessly asphyxiated in a gas chamber as a humanitarian means rather than exposing them to the miseries of the foundling hospitals.

Even in modern society, infanticide is far from unknown. Corpses of newborn infants are sometimes found on the screens of city sewage-disposal plants —a method of disposing of unwanted children which has been practiced for thousands of years. Also the widespread incidence of child abuse is beginning to crawl out of the woodwork into the light of day. At the 1973 Congressional hearings on the Child Abuse Prevention Act, it was estimated that at least 60,000 children die annually in the United States from child abuse, with severe neglect and/or actual abuse affecting one out of every 100 children at this time.

The next most common positive check to population growth has been induced abortion. Abortion has been widely used in practically all cultures at all times. In a study of 400 primitive societies prepared by George Devereux, only one group could be found in which induced abortion did not occur. Numerous methods were reported, such as the use of certain drugs, deliberate starvation, hard labor involving movement of the pelvis, lifting heavy objects, jumping from high places, use of tight belts and girdles, and placing heavy weights on the abdomen.

There is no question that abortion has also been very common in Western cultures through the years, although it was not considered a suitable subject for discussion, and little has appeared in print. It is no secret, however, that ecclesiastical authorities have long been concerned about the extent of abortion, as witness one injunction of an early midwife's oath: "You shall not give any counsel, or minister any herb, medicine, or potion, or any other thing, to any woman being with child, whereby she should destroy or cast out that she goeth withal before her time."

In the past fifty years, abortion has been legalized in many European countries including Poland, Hungary, Rumania, the Soviet Union, and the Scandinavian countries. Abortion rates have sometimes reached astonishing levels; for example in 1934 Moscow reported 2.7 abortions for each live birth. In the 1960s the abortion rate in Hungary exceeded live births, and a similar situation existed in Japan while it was desperately trying to restrain population growth after World War II. Even in countries where abortion is illegal, the rate is often unbelievably high; for example, it has been estimated that the abortion rate in Catholic Italy is equal to the live birth rate—probably a direct result of the limited availability of information on birth control. A 1971 survey estimated that, on a worldwide basis, there were more than four legal and illegal abortions for every ten live births.

In 1955 at a conference of the Planned Parenthood Federation of America, it was estimated that there were half a million illegal abortions yearly in this country. Even our best hospitals quietly got into the act, and one of the most commonly performed surgical procedures was the "D and C" (dilation and curettage), in many cases a cover-up for abortion.

In 1973, the United States Supreme Court held that during the first three months of pregnancy the abortion decision must be left to the medical judgment of the pregnant woman's physician. This led to a rapid rise in the abortion rate, which in seven years has more than doubled to today's value of about 1.5 million per year. Thus there is no question that abortion has become a significant means for population control.

A properly performed abortion involves less risk than normal childbirth. Many authoritis such as Garrett Hardin point to abortion as a very important "backstop" to be used when other methods have failed. While the "pill" is about 99 percent effective, even this small failure rate is producing a quarter of a million unwanted pregnancies in the United States each year, and many abortions are being performed on married women who do not want more children. For the incredible number of unmarried teenagers who become pregnant abortion is often considered the best solution.

Abortion is now coming under considerable pressure by "Right to Life" advocates who correctly note that abortion is now commonly used as a substitute for other methods of contraception. There is also the question as to whether the public should pay for "welfare" abortions. One thing is obvious: those for and against abortion are so polarized in their views that any compromise is impossible. Strong attempts are now being made to once again outlaw abortion altogether, but it is hard to see how such a law could be effective. As it has in the past, abortion would simply go underground with all the attendent evils of such a system.

Preventive Checks

Birth control* in one form or another has been practiced throughout all recorded human history. Contraception was described in Egypt as early as 1850 B.C. and possibly even earlier in China. Primitive societies tried to control conception by sex taboos, limiting the time and frequency of intercourse, various perversions, and prolonged lactation. The Greeks and Romans were thoroughly familiar with gynecology and contraception, and several Roman books were unequalled for almost 2000 years. The story is perhaps best described by Dr. Norman Hines in his *Medical History of Contraception:*

> Contraception has existed in some form for at least several thousand years. The *desire for,* as distinct from the *achievement of,* reliable contraception has been characteristic of many societies widely removed in time and place. This desire for controlled reproduction characterizes even those societies dominated by mores and religious codes demanding that people increase and multiply. Men and women have always longed for both fertility and sterility, each at its appointed time and in its chosen circumstances. This has been a universal aim, whether people have been conscious of it or not. However, only since 1822 has there been any organized planned effort to help the masses to acquire a knowledge of contraception.

Malthus did *not* believe in birth control, which he considered unnatural and immoral. His "preventive" or conception-limiting checks consisted primarily of late marriage and of abstinence within marriage, namely:

> . . .A foresight of the difficulties attending the rearing of a family acts as a preventive check. . . and appears to operate in some degree through all the ranks of society in England. There are some men, even in the highest rank, who are prevented from marrying by the idea of the expenses that they must retrench, and the fancied pleasures that they must deprive themselves of. . . The labourer who earns eighteen pence a day and lives with some degree of comfort as a single man, will hesitate a little before he divides that pittance among four or five, which seems to be but just sufficient for one. . . .

It is impossible to believe that Malthus did not know that birth control was being widely practiced in Europe during his lifetime. For example, the birth rate in France had been decreasing since the middle of the eighteenth century, largely through the practice of *coitus interruptus* (withdrawal). By the nine-

*Theoretically, there is only one kind of *birth* control—abortion; methods such as the "pill," spermicides, and mechanical devices are really examples of *conception* control.

teenth century this had spread to the Scandinavian countries, and starting about 1850 there was a declining birth rate for all of Western Europe.

The modern birth-control movement goes back to Francis Place (1771–1854), the son of Simon Place, "an energetic but dissipated man" who was the keeper of a sponging house in London. As a boy, Francis was often beaten by his perennially drunken father who also frequently deserted his family. In 1791 at the age of 19, Francis married Elizabeth Chadd, a young woman who "proved the great moral influence of his life, and lifted him from the mire of his past surroundings." Marriage also brought tremendous economic burdens, including the advent of a labor strike which left them on the verge of starvation for eight months during which he could find no work and the eventual birth of fifteen children, five of whom died in childhood.

Overcoming his humble origin, Francis Place worked himself upward to the profession of master tailor so that by the age of 50 he was already retired with a comfortable income and deeply concerned with the social problems of the day. His own early marriage had been his salvation, and the Malthusian recommendation of late marriage and moral restraint appeared to him an utter absurdity. Place himself had failed to "live decently in celibacy" even to the age of 19, and the idea of a typical laborer waiting to take a wife until he had an assured means of support seemed only a recipe for hopeless immorality. Thus Place became obsessed with the feeling that Malthus' remedies were impractical, and in 1822 his *Illustrations and Proofs of the Principle of Population* was published, the first book in English to propose contraception as a substitute for Malthus' moral restraints.

Even more important was Place's distribution of contraceptive handbills among the working classes. These received considerable circulation and publicity not only in London but also in the industrial towns of the North. Addressed to "The Married of both Sexes," they contained the following advice:

> A piece of soft sponge about the size of a small ball attached to a very narrow ribbon, and slightly moistened (when convenient) is introduced previous to sexual intercourse, and is afterwards withdrawn, and thus by an easy, simple, cleanly and not indelicate method, no ways injurious to health, not only may much unhappiness and many miseries be prevented, but benefits to an incalculable amount be conferred on society.

The American birth-control movement began about 1828 and owed its origin partly to the work of Place and others in England. The most significant contribution was the publication in 1832 of *Fruits of Philosophy,* by Dr. Charles Knowlton, a small-town physician in Western Massachusetts. Knowlton's book was an excellent treatise on gynecology and contained much information commonly found in today's sex manuals. Chapter III, entitled "Preventing conception without sacrifice of enjoyment," contained Knowlton's recommended method of douching with an astringent solution such as aluminum sulfate, zinc sulfate, sodium bicarbonate, or vinegar. Knowlton's book was widely distributed, and he was soon involved in a series of legal entanglements in puritan Massachusetts. In 1832, he was fined at Taunton, Massachusetts, and in 1833 he was jailed for three months in Cambridge for distributing his book. All of this publicity made the book internationally famous, and as many as half a million copies were eventually sold.

In 1873 the United States Congress passed the notorious Comstock Law, which made it a criminal offense to import, mail, or transport in interstate commerce "any article of medicine for the prevention of conception or for causing abortion." Anthony Comstock headed the Society for the Suppression of Vice at the time, and his efforts led to enactment of the law. The Comstock Law also applied to "obscene literature," which included any description of contraceptive devices or methods. By the start of the twentieth century, however, more effective methods of contraception control had been devised, and courageous workers such as nurse Margaret Sanger had opened the first birth-control clinics. In spite of localized pockets of resistance, legal opposition to birth control in the United States was essentially dead.

Was Malthus right?

In spite of the violent personal attacks on him, it is generally agreed that Malthus was gentle and good-natured, "one of the most serene and cheerful of men." He was a devoted family man and a faithful friend even to those who strongly disagreed with him. He was a teacher of high integrity and was well liked by his students. In his own words, "My ultimate object is to diminish vice and misery."

Was he right in his predictions? As far as nineteenth-century England was concerned, Malthus could not have been more wrong. The industrial revolution combined with improvements in agricultural methods, better means of transportation, improved public health, and overseas trade with Canada, Australia, and New Zealand all served to bring about a rapid rise in the British standard of living. Population grew rapidly both at home and throughout the Empire, and at the start of the twentieth century the British truly "ruled the waves."

When we consider many other parts of the world, however, Malthus not only was but is still correct. For generations, population growth in India, China, Bangladesh, and many African nations was limited by the Malthusian checks of malnutrition, disease, and periodic famine. Today it is becoming increasingly obvious that there are limitations on the ability of the world to produce food. Unless worldwide population growth can be stopped, it is only a question of time before the entire world experiences a Malthusian disaster.

There is no question that Malthus has been subject to much unfair criticism in the years since the publication of his *Essay.* Many of his most severe critics never read his works, and the word *Malthusian* became synonymous with vice, misery, and starvation. Perhaps a fairer account of his contributions is provided by William Peterson, who wrote in the magazine *Demography:*

> Is this word (Malthusian) ever used to designate, say, the first significant economist to recognize the importance of effective demand and thus the only nineteenth century figure in the main line of classical economic thought to suggest the serious lacks in laissez-faire policies; or in social thought, a pioneer advocate of universal education, the initiator of political science as a university discipline; or, specifically with respect to population, the theorist who analyzed both the relation between humans and the effect of social man's rising aspirations on his fertility? Very little of the full and well rounded thought of Professor Thomas Robert Malthus is recalled in the commentary even of professionals.

A Matter of Life and Death

It is absolutely essential that developed and undeveloped nations alike recognize the magnitude of the population problem. At their present rate of growth, the undeveloped nations are contributing an additional 75 million inhabitants each year to spaceship earth. This explosive growth has soaked up any increase in resources made available to them in recent years; in fact, many of these countries now face worse conditions than they did thirty years ago when the massive aid programs were begun. While additional food supplies can alleviate periodic shortages and prevent starvation for a time, we are learning that this assistance only adds to the potential for an even greater disaster at some future time.

In addition, it has become obvious that increased immigration by the developed nations can absorb only a small fraction of the world population growth. In the summer of 1979, many of us were horrified to read Malaysian statements that refugees from Vietnam would no longer be accepted, that the "boat people" would be driven back to sea and, if necessary, their boats would be sunk!

Although the idea of drowning refugees is repugnant to most people in the world today, it may unfortunately be a foretaste of future attitudes. Each immigrant places an additional load on the resources of his new homeland, thereby lowering the standard of living for all of its inhabitants. It is entirely possible that immigration will be essentially eliminated throughout the world in the not too distant future, and no nation will be permitted to "export" its population problem.

The world common is getting pretty crowded. Pollution is an increasing problem, families everywhere are being asked to think twice before having more children, and the developed nations are finding it necessary to reduce consumption of energy and raw materials. The increase in world population during the decade of the 1960s was about the same as that for the entire nineteenth century. At the present rate of population growth, each inhabitant of the world will have some 20 percent less living space, water, recreation, and reserves of fuel and minerals available to him every ten years. If voluntary methods of birth control continue to be rejected by a large fraction of mankind, far more drastic measures may prove to be the only means of survival for future generations.

Chapter Three

Our Daily Bread

We have seen that prehistoric man was dependent on hunting for his food, with all the dangers and the uncertainties as to availability of game. The Neolithic revolution made possible a considerable expansion of food supplies through agriculture and thereby overcame the vicissitudes of the hunt; even then, however, there was always the possibility of crop failures due to weather, insects, or warfare, and periodic shortages were undoubtedly a way of life.

There was one saving grace in the precarious existence of these early farmers: when famine occurred, it was largely localized and often confined to only a small tribe or village; those living only a few miles away could be relatively unaffected, so that mass starvation was much less likely. This whole situation changed with the development of the Greek and Roman empires, particularly with the growth of their larger cities. Early Athens is believed to have had a population of several hundred thousand, and Rome may have approached a million inhabitants. Survival of these cities was dependent on the availability of surplus food from the surrounding countryside, and transportation facilities were quite limited. Crop failures could affect tens or hundreds of thousands of people, and the stage was set for mass disaster.

For those of us who are fortunate enough to live in the developed nations today, it is difficult to have any real understanding of famine. Not only do

Figure 8. Farming in India: Men throw grain into the air to separate the wheat from the chaff, a woman drives oxen through grain-stalks to break it into "fodder," while the farm overseer contentedly puffs on his hookah. (World Wide Photos, Inc.)

human beings suffer and die in large numbers, but many are driven to extreme behavior by the stress of starvation. In addition to stealing and hoarding food, starving people will eat clay, vermin, and pulverized bones; they will sell children for money and will even resort to murder, cannibalism, and suicide. Some idea of the frequency of famine can be obtained from a book written in 1878 by an Englishman, Cornelius Walford, which describes 350 famines of record. Most were small localized famines, but a few were true disasters and accounted for millions of deaths.

Walford writes, for example, that in 436 B.C. thousands of Romans allegedly threw themselves in the Tiber River rather than slowly starve to death. In 310 A.D., famine killed 40,000 persons in England. For seven years in the eleventh century, the annual overflow of the Nile River in Egypt failed—and with it almost the entire subsistence of the country. At that time the wretched inhabitants resorted to cannibalism, and organized bands kidnapped unwary travelers in the desolate streets. Between 1193 and 1196, incessant rains in England and France destroyed food crops, and the common people perished everywhere for lack of food.

In 1600, famine combined with plague in Russia took half a million lives. Even as late as 1920–21, millions of Russians died as a result of drought, and cemeteries in some areas were guarded to prevent the starving people from digging up freshly buried corpses.

Probably the best known European famine occurred in Ireland in 1846. The potato had been brought to Spain and Portugal from South America in about

1565; it spread gradually throughout Europe and reached Ireland by the eighteenth century. Plagued by war and civil disruption, the Irish decided to adopt the underground potato as a staple crop to protect farmers from recurrent burning and destruction by the English. This highly successful crop greatly expanded the food supply, and the Irish population grew rapidly from slightly over one million in 1690 to over eight million by the middle of the nineteenth century. In the summer of 1846, potato blight struck the country, and the resulting famine of 1846–47 killed about 1.5 million people. Large-scale emigration to the United States further reduced the population; since then, delayed marriage and low birth rates have maintained the Irish population at about four million, only half what it was before the potato blight struck.

In terms of frequency and mortality, however, famine in Europe is nothing in comparison to famine in Asia. Much of Asia's agriculture is dependent on the annual monsoon rains which sometimes fail altogether and sometimes come in excess. The vital rice-growing areas of India, China, Pakistan, and Bangladesh, which support the world's largest and densest population, are particularly susceptible to any change in the monsoons and have been the sites of the world's worst known famines.

The first great Indian famine of record was in 1769–70, when drought in Bengal caused as many as ten million deaths. As Walford describes it, ". . . The air was so infected by the noxious effluvia of dead bodies, that it was scarcely possible to stir abroad without perceiving it; and without hearing also the frantic cries of the victims of famine who were seen at every stage of suffering and death. . . ." The next 200 years saw at least another ten million deaths by starvation in India. During the war years of 1943–44, excessive rains destroyed much of the rice crop and difficulties of supply caused two to four million deaths in Bengal alone.

Things were just as bad in China. There was almost no rain for three years between 1877 and 1879 in North China, and from nine to thirteen million died of starvation. People were openly sold in the market for food—a respectable married woman could be bought for $6 and a little girl for $2. Famines in 1920–21 and again in 1929 caused an additional 2.5 million deaths in China. Even during more or less normal years, deaths from starvation were estimated to be over a hundred thousand a year well into the twentieth century.

Since World War II, the availability of surplus grain from the United States, together with improved means of global distribution, has made possible rapid delivery of large amounts of food to any point on the globe. Under these conditions, the possibility of any large-scale famine due to natural causes is very remote. The much-publicized Biafran famine of 1969–70 in Nigeria resulted in several hundred thousand deaths, but in this case foreign aid was prevented by military operations.

PHOTOSYNTHESIS

All life on earth is dependent on the process of *photosynthesis,* whereby green plants are able to use part of the solar energy that reaches the earth's surface to produce the large organic molecules such as starch, sugar, cellulose, lignin, and protein, which are essential for life. An important by-product of the photosynthesis reaction is oxygen, which is returned to the atmosphere. The essential requirements for photosynthesis are: (1) *sunlight,* (2) *carbon dioxide,* (3) *water,* and (4) *minerals,* particularly nitrogen, phosphorus, and potassium (legumes are an exception, since they utilize nitrogen directly from the atmosphere).

In addition to providing the food essential for animal life, photosynthesis makes possible the *carbon cycle,* which involves the recycling of oxygen to and from our atmosphere. When animals consume plant material, they use oxygen from the air to burn their food to carbon dioxide and water. Plant material not consumed by animals eventually dies and is also converted back to carbon dioxide and water by microorganisms in soil or in water. The carbon dioxide released to the atmosphere is again available for photosynthesis, which returns oxygen to the atmosphere and thereby completes the cycle.

Estimates of the total yearly production of plant life (*biomass*) throughout the world are more than 100 times the recommended daily calorie requirements for the entire world population. Why is there a food problem in the world today? The fact is that over a third of the photosynthesis occurs in the ocean, where a very complicated food chain leaves little that mankind can use, although a good deal of research has been done on the subject. In addition, most of the photosynthesis occurring on land produces cellulose in the form of forests, shrubland, and grass, and (unlike the goat) *Homo sapiens* is unable to digest and utilize cellulose. More obvious to us is the large fraction of undigestible material—such as roots, stems, and leaves of our common food plants—produced by photosynthesis occurring on cultivated lands.

REQUIREMENTS FOR GOOD NUTRITION

It has been very hard to establish minimum food requirements since these vary greatly from one individual to another. Some persons are just more efficient in utilizing their food intake and thus require less. Manual laborers require more than office workers, women (in general) require less than men, and young children require less than adults. It is certain that pregnant women, nursing mothers, and teenagers require more than average food intake. It is also commonly assumed that residents of hot countries require less food than those in colder climates, although this concept has been challenged recently.

For the average adult, roughly 1700 calories a day are required for bare existence. The typical resident of India or Africa needs some 2300 calories per day, and this is increased to about 2700 calories per day for the typical resident of Canada or the Soviet Union. The United Nations Food and Agriculture Organization bases many of its calculations on a population-averaged value of 2350 calories per day, which is considered to be a generous amount.

Even more important than the total calorie content of food is the amount of available protein. This is particularly serious for young children from ages 1 to 6, where a protein deficiency can harm physical and mental development by impairing the growth of the brain and central nervous system. Protein deficiency is also particularly harmful for pregnant women and nursing mothers and their children; even more protein than normal is required at this time.

The recommended daily requirement for protein is often given as 1 gram per kilogram of body weight. For an average adult, this corresponds to about 60 grams per day of total protein, including for good nutrition at least 10 grams per day of animal protein; again, this is considered to be a generous amount. Most people who suffer from a shortage of calories will also be found to suffer protein malnutrition, and there are some parts of the world where starchy foods supply sufficient calories but insufficient protein.

Proteins are composed of some twenty different amino acids, of which eight are essential for human nutrition. Meat, fish, eggs, and milk each contain all the amino acids required by the human body and are known as *complete proteins.* All vegetables and cereals, including rice, wheat, and corn, contain incomplete protein, which is lacking one or more of the essential amino acids. Legumes (including most nuts) are also incomplete, although these are quite good sources of protein. It should be noted also that, even though each individual component is incomplete, it is possible to combine different foods such as peanut butter and whole wheat bread to produce a high quality balanced protein.

Vitamins and minerals are also essential for life, and sufficient amounts are usually obtained in the average American diet. Eggs, liver, dairy products, as well as some fruits and vegetables are important sources of Vitamins A and D. Whole grains, liver, and yeast supply the Vitamin B complex, and Vitamin C (ascorbic acid) is found in fruits and leafy vegetables. About the only minerals that sometimes present a problem in the American diet are calcium and iron.

It cannot be emphasized too strongly that a balanced diet of wholesome food will meet all nutritional needs, and that a dietary supplement is seldom necessary. The human body is remarkably adaptable and can do very well without daily vitamin capsules, instant fortified breakfast, and a certain number of ounces of orange juice. For many centuries, northern Europeans got along quite well on a diet of bread, potatoes, cabbage, and small amounts of meat.

The essential thing for good nutrition is to eat a variety of foods in moderation—which means avoiding fad diets, such as the combination of wheat germ, yogurt, and blackstrap molasses recommended some years ago for youthful longevity! More recently, there have been the Zen macrobiotic diet, the grapefruit diet, and several varieties of vegetarian diet. Diets such as these limit the natural choice of food, and it is often difficult to insure that all the essential food elements are present.

The greatest food problem in the United States is overeating. At any given time, roughly 12 percent of our population are dieting to lose weight, 20 percent are watching their weight, and another 10 percent are wondering if they should be watching their weight. Americans spend some $10 billion each year trying to lose weight; 95 percent of those trying either fail to lose the unwanted pounds or else cannot keep them off. American psychiatrist Henry A. Jordan told food editors at a 1980 conference, "Obesity is the number one health problem in the country."

Much of this problem, of course, is the direct result of a craving for high-calorie junk food which is being fostered in us from early childhood; the sad fact is that many people are as strongly addicted to foods of low nutritive value, junk food, as others are to smoking or drugs. It is certainly significant that, in the 1970s decade alone, fast-food chains reported a five-fold increase in their business, and about half of the population of the United States now eat a majority of their meals away from home. It is easy to consume 800 calories in 10 minutes at a fast-food restaurant, and it is seldom the type of food that gives real satisfaction or "sticks to the ribs." An hour or two later, the food junkie feels hungry again and puts away a regular meal (or perhaps another snack) without giving thought to the hundreds of calories involved or the probable lack of important nutrients.

WORLD FOOD PRODUCTION

On a worldwide basis, rice and wheat are the two most important food crops, each accounting for about one-fifth of our food energy. The third most important cereal crop is corn, which is widely used for human consumption in Latin America and Africa; most of the corn grown in the United States and Europe is used for animal feed. Livestock and fish contribute about one-tenth of the world's food energy and are important sources of protein. Potatoes, vegetables, fruits, nuts, sugar, fats, and oils make up the balance of our diet.

Since the time of the industrial revolution, consistent and increasingly rapid world population growth has placed increasing demands on our food supply. There are two ways by which the world food supply can be increased: bringing more land into cultivation or raising the *yield* of land already in cultivation. From the Neolithic revolution through World War II, most of the growth came from expansion of the cultivated area with very little change in the yield per acre. Development of agriculture by Greece and Rome, cutting down of the great forests of Europe, discovery of the New World, and the opening up of the great land areas of the United States, Canada, Argentina, Australia, and New Zealand each greatly added to world food supplies. Most of the suitable land has now been brought under cultivation, and increases in world food supplies since World War II have come primarily from increased productivity of land already under cultivation. During the fourteen years between 1963 and 1977, for example, the world's cultivated land area was increased by only 6 percent while its total food supply increased by over 40 percent—at the same time the world population increased by more than 30 percent. It now appears that any possible increases in our future food supply will have to come by this route.

Land Productivity

The spectacular increase in land productivity since the 1940s has come about primarily as a result of a better understanding of plant genetics—with a further contribution from increasing use of fertilizers and pesticides. By selecting homogeneous strains of seed stock, it is possible to make plants shorter or taller, more resistant to disease, more tolerant of cold, and more responsive to fertilizer. The northern limit of commercial corn production in North America has now been extended 500 miles, and varieties of wheat have been developed that give good yields even near the equator. Most important, yield per acre has been greatly increased; for example, in the United States the corn yield has more than doubled with the introduction of *hybrid corn,* and similar increases in the yields of wheat and rice have been achieved in many parts of the world.

Food production in the undeveloped countries also increased following World War II, but—as we have seen—population growth soaked up most of this increase. During the 1960s, food production per person was actually declining in many areas, and worldwide famine loomed as a real possibility. Asia, Africa, and Latin America had been grain exporters prior to World War II; by the 1960s, they were importing millions of tons a year. At the same time, the "planned economies" of the communist nations were not meeting their goals. China, the Soviet Union, and even East Germany (once the breadbasket of all of Germany) became food importers. When the monsoons failed for two successive years in 1966–67, only huge imports from the United States (including 7 million tons of wheat) prevented millions of deaths by starvation in Southeast Asia.

In the late 1960s, a concerted effort was made to improve the grain yields of the undeveloped countries. Known as the *green revolution,* this involved the introduction of high-yield dwarf varieties of wheat and rice which permitted far more efficient use of agricultural resources. The traditional wheat grown in Asia had been characterized by a tall thin stalk which often fell over when too much fertilizer was applied. This caused severe crop losses and thus limited the use of fertilizer as a means for improving yield. By substituting a short dwarf variety, increased application of fertilizer and water could easily double the yield of the previous varieties. In addition, the dwarf varieties were remarkably adaptable to different growing conditions, permitting them to be used at latitudes ranging from the equator through the temperate zone.

Similar success was obtained in the case of rice when some 10,000 different varieties from all over the world were assembled for crossbreeding. When a tall vigorous variety from Indonesia was crossed with a dwarf rice from Taiwan, the first *miracle rice* was obtained; with proper use, this could double the yield of most traditional rices grown in Asia.

Introduction of new varieties of rice and wheat has postponed—at least temporarily—any severe worldwide catastrophe. In fact, from the index of worldwide food production per capita, it can be seen that there has actually

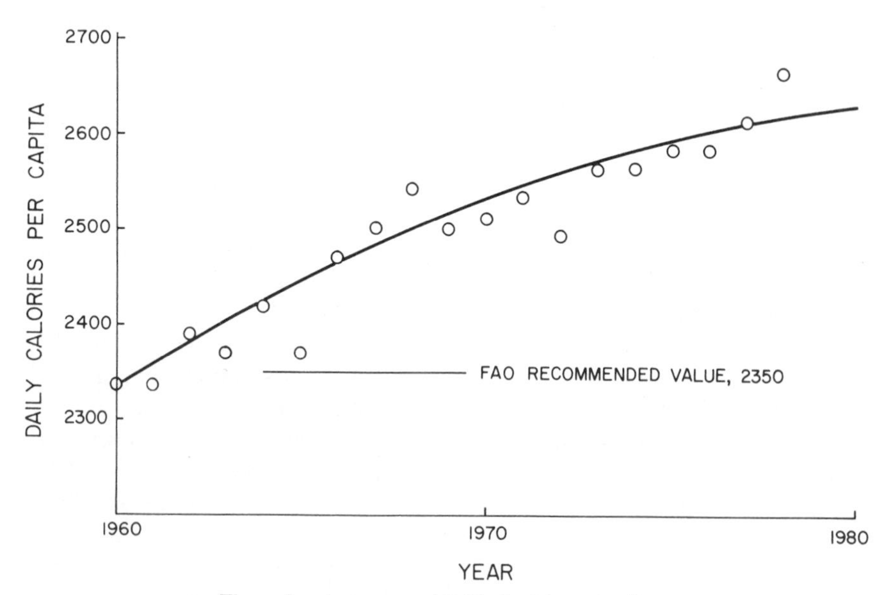

Figure 9. Average worldwide food consumption.

been about a 10 percent increase during the past twenty years. Today there are approximately 2600 calories (see Figure 9) including 70 grams of protein, available daily for the entire world population—according to the figures.

So What is the Problem?

Since we know that the recommended food intake per person is 2350 calories, including 60 grams of protein, the answer obviously lies in the fact that these food resources are not distributed uniformly. The developed nations consume appreciably more than the world average, which means the others have less. At present, the average individual in Africa has only some 2200 calories per day, and food production per capita has decreased substantially over the past ten years because of rapid population growth. Food production in Asia has barely kept up with population growth (see Figure 10). India, the world's second most populated country, is barely maintaining a level of about 2000 calories *per capita* with an inadequate supply of protein. China is in pretty good shape, and Latin America as a whole is also relatively well-off, although some nations, such as Bolivia and Ecuador, are marginal.

Not all food produced, of course, is directly available for human consump-

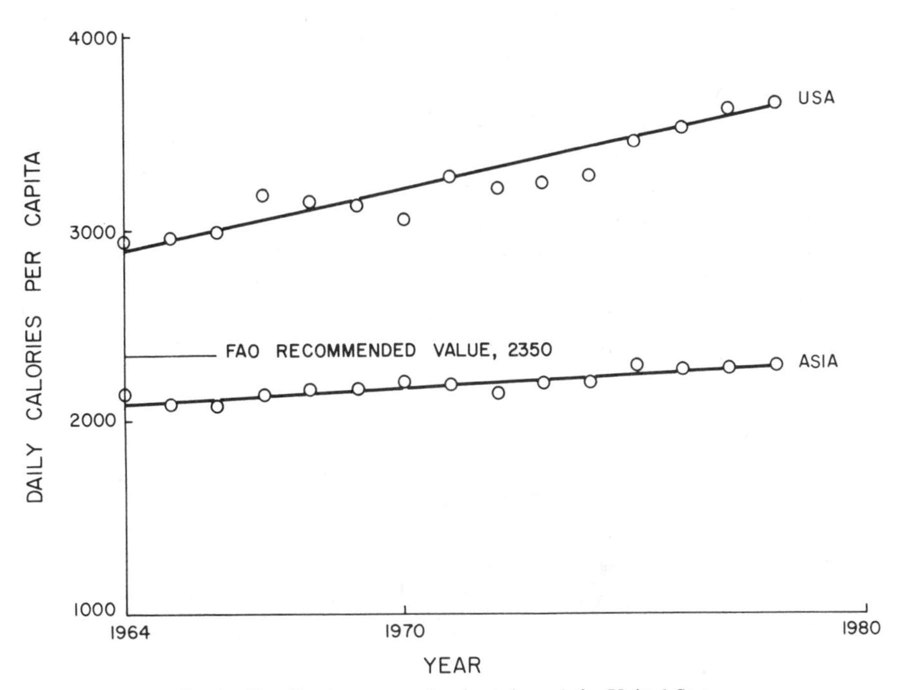

Figure 10. Food consumption in Asia and the United States.

tion. The United States produces an equivalent of about 10,000 *original daily calories* per capita, but most of this is fed to animals and appears as meat, eggs, and dairy products. Daily consumption per capita in the United States is about 3500 calories, with some 40 percent of these calories coming from animal products. In Asia, for comparison, less than 5 percent of the total calorie intake comes from animal products. Our daily consumption of protein is over 100 grams. The 168 grams of fat consumed daily by the average American is almost three times the world average, and animal products that are high in cholesterol account for over 70 percent of the fat in our diet.

Consumption of meat in the United States is four times the world average. Particularly disturbing is our consumption of beef, which has almost doubled since World War II—although it has begun to decline in recent years because of the high prices. This increased consumption of beef is probably a direct result of the tremendous growth of fast-food outlets with their continuing emphasis on hamburgers. Poultry consumption has tripled since World War II because of increased production efficiency and relatively low prices compared with other meats. Pork consumption has remained fairly constant. Consumption of both milk and eggs has dropped appreciably in recent years because of lower-calorie diets and warnings about cholesterol.

In addition to our excessive dependence on animal products, Americans are also the largest consumers of junk food. We each consume on the average 100 pounds of sugar per year, which contributes some 15 percent of our total calorie intake. This includes several billion pounds of candy sold each year. A recent government survey reported that 40 percent of the population seldom or never eat cereal but only 13 percent do not eat desserts. Over one-third admit to regular munching of salty snacks, and one-fifth drink five or more cups of coffee each day. Our yearly consumption of soft drinks is 36 gallons per capita, and our beer consumption is 22 gallons per capita. We consume 38 million pounds of synthetic vitamins each year—enough to supply 300 capsules to each man, woman, and child. It is not surprising that one-third of the males and one-half of the females over the age of 16 feel they are overweight.

FUTURE PROSPECTS

Although malnutrition is common in many regions because of the nonuniform availability of food, the world does produce enough food today to supply an adequate diet for each of its inhabitants. Even with the tremendous surge in population, there has been a slight increase in world food production per inhabitant. Despite these facts, however, many international experts are concerned about the possibility of an eventual major food shortage leading to mass starvation on a scale never seen before. What is the basis for this view?

The foremost problem is undoubtedly the rapid population growth of the undeveloped nations, which have continued to grow at an essentially constant rate of 2.3 percent a year for more than thirty years. Furthermore, as we have seen in Chapter 2, any attempts to persuade these governments to adopt population policies are usually interpreted as self-serving "imperialist" or "neocolonial" policies on the part of the developed nations. Established United Nations doctrine is that each nation has absolute sovereignty with respect to its population policies, and ultimate decisions are made at the level of the family with full freedom to decide on the "number and spacing of children." The common attitude was perhaps best expressed by an African woman delegate to the 1974 United Nations World Population Conference, who said, "You want us to go back to our villages and take your pills. You overintellectualize everything. Why don't you listen to us for a change?"

The second major problem is that the ultimate capacity of the world to produce food is limited and that we may be now approaching that limit. The green revolution has staved off mass starvation for a time, but it is unlikely that there will be any new biological breakthroughs that will lead to such spectacular increases in yield per acre. Also, we have learned that the improved crop varieties often have severe disadvantages. The higher yields are dependent on adequate supplies of water and fertilizer, two things that are becoming less available and more expensive each year. There is still some question whether the new varieties are more susceptible to disease (hence requiring increased use of costly insecticides). For example, several years ago most of the hybrid corn crop in Iowa was badly hit by a wilt which substantially decreased the yield. With a large fraction of the world's food production now based on seed that is genetically homogeneous, it is entirely possible that a new plant disease may come along some day and wipe out a large fraction of the *world* food crop.

It is also very unlikely that the cultivated land areas of the world can be substantially increased in the future. There are just no more "New Worlds" to discover. Other than ice, desert, tundra, and similar regions unsuited for agriculture, the only large uncultivated areas left are the tropical rain forests of Africa and South America, and the farming of tropical soil presents many real problems. Most of the available nutrients occur near the surface and in the dense vegetation now growing on such soils; when the land is cleared, the heavy rainfall rapidly leaches out the few nutrients left in the soil so that after two or three crops the soil loses its fertility. In addition, the high temperatures often cause the soil to compact into a hard material known as *caliche,* which can no longer be plowed. The African rain forest is the home of the *tsetse* fly, carrier of sleeping sickness and the cattle-killing disease *trypanosomiasis.* With this insect endemic in the wildlife, efforts at agriculture have been easily discouraged. Furthermore, we are now aware that extensive clearing of the rain forests might have a disastrous effect on world climate, which would more than offset the benefits of any small agricultural production.

Centuries of abuse of the land by overgrazing, deforestation, burning, and poor agricultural practices have changed many of the formerly fertile regions of the world into wastelands. The Mediterranean Sea was once surrounded by wooded, fertile lands. North Africa was the granary of the Roman Empire. In the middle of the great coastal plain of Tunisia today, ruins of the Roman city of Thydrus include a coliseum that seated 60,000 spectators and was second only to the coliseum of Rome. From supporting an extensive irrigated agriculture of grain fields and olive orchards, this area has become largely desert, able to support only a few grazing animals.

Destruction of productive land continues unabated today. Overgrazing plus several years of drought in the Sahelian region of Africa have brought severe famine and are extending the Sahara Desert by as much as 30 miles a year in the worst areas. Millions of acres of cropland in food-deficient parts of Asia, Africa, and the Middle East have been abandoned in recent years because of lowered productivity resulting from soil erosion. Deforestation of the Himalayas and surrounding foothills has caused severe flooding of major river systems, such as the Indus, Ganges, and Brahmaputra, with much soil erosion and resulting loss of farmlands in India and Bangladesh. Even the United States has been guilty of overcultivation, as witness the Dust Bowl of the 1930s and the current problems of increasing salinity in important irrigated regions such as the Imperial Valley of California.

Farming the Oceans

Since over one-third of the world's photosynthesis takes place in the oceans, it was hoped at one time that these "immeasurable riches" would go a long way toward the solution of our food problems. It now appears, however, that our oceans can contribute very little beyond their present yield. Over 90 percent of the ocean is open sea, which lacks the dissolved nutrients needed for high yield of marine life. Practically all of the world fish catch occurs close to shore and in certain coastal areas where upswelling cold currents bring nutrients to the surface, and even these areas have a limited yield because of the complicated food chain. Photosynthesis actually occurs in very small marine organisms known as *phytoplankton,* which are present in large numbers throughout the oceans and serve as food for small marine animals. "Big fish eats little fish" until eventually there is one big enough for man to enjoy. Each step in this chain involves considerable loss, however, so that only a very small fraction of the phytoplankton can be utilized for man's benefit.

Between 1950 and 1970, the world fish catch increased about 5 percent each year. This increase was largely the result of technological advances in the fishing industry. Sonar equipment was developed to better locate schools of fish and, by 1969, was so successful that the British east coast herring industry was almost wiped out because of virtual destruction of the breeding stock. In New Zealand waters, a single Rumanian "factory ship" caught as much fish in only 1 day as the entire New Zealand fishing fleet of 1500 vessels. In United States waters, more efficient foreign fishing fleets in 1973 caught over five times as much fish as the entire United States fleet of smaller vessels. Whaling, too, has been so successful that several species have been pushed close to extinction, and many authorities believe that all whales will soon disappear unless strict international controls are imposed.

The harvesting of fish from our oceans presents a perfect example of the "tragedy of the commons." For many years, it was assumed that the oceans of the world were free to anyone beyond the generally accepted 3–mile territorial limit. Under these terms, nations such as Japan and the Soviet Union have ruthlessly plundered the world's best fishing areas with no concern for the protection of future supplies. Only when it became apparent that this was having a disastrous effect on the supply of important breeding stocks did the nations most concerned arbitrarily extend their fishing rights to a 200–mile limit. This represents a step toward adequate management of the world's fishing areas and, it is hoped, toward management of one of its major sources of food.

In 1971, the world fish catch reached a value of about 80 million tons; the following year it declined, and since then it has fluctuated between 75 and 82 million tons. It has been estimated that the maximum sustainable yield with good management worldwide is about 110 million tons, only one-third more than today's catch. With an increasing world population and greater pollution of important fishing areas, it is apparent that fish will contribute a smaller fraction of the world's food needs in years to come.

Although the fish catch represents only about 2 percent of the total world food, it is very important because it supplies a large fraction of the world's animal protein, either directly as food or indirectly as livestock feed. Some nations, such as Japan, depend strongly on fish as a source of protein and, in fact, their per capita consumption of fish is far greater than that of meat. For the Japanese, the loss of even a fraction of their fish catch would mean certain disaster.

One way of increasing the annual fish production would be to go down the food chain from big fish to small fish or even to the lowly phytoplankton. For example, Peru has harvested tremendous quantities of anchovies for the production of high-protein fish meal, most of which has been used in poultry feeds. Millions of tons of small shrimp-like krill could be harvested in the Antarctic and processed to a palatable food for humans. It is also possible to increase fish production by fish "farming" in either fresh or salt-water ponds. This has been done for thousands of years in the Far East and is still common in China and Indonesia for the raising of carp and other species of fish. In the United States, the possibility of raising shrimp and lobster in special ponds has been explored along with the idea of utilizing waste heat from electric power plants for steam-heated fish farms.

Alternative Solutions

The most obvious means of increasing the amount of food available to the undernourished people of the world is "conservation" (in other words, *eating less*) by the overly nourished. There is something badly out of focus when a large fraction of the population is not only overweight but also unhappy about it. The overfed Americans are not alone in this; it is a problem common to all developed nations. Emphasis on improved nutrition for all, combined with moderation of eating habits in these nations, could make available tremendous quantities of food—and simultaneously control one of today's major health hazards, obesity.

The United States, in particular, could conserve food by gradually shifting away from our present strong dependence on animal products. If we were to consume more whole-grain cereals and reduce our consumption of meat by one-half, this (combined with a modest reduction of calorie intake) would lead to a reduction of our present average use of 10,000 *original* calories per person by about one-half. Lowering consumption of these saturated fats would, of course, be very beneficial to our arteries as well.

It has also been suggested that a reduction of our ever-increasing pet population could also save considerable amounts of high protein food. Pet food sales in the United States alone are now more than $3 billion annually, and it is no exaggeration to say that the average pet enjoys better nutrition than the average human being. The high-quality protein (fish meal, horse meat, and animal by-products) that presently goes into pet food would go a long way toward supplying the protein needs of a nation such as Bangladesh.

Considerable work is currently being directed toward the development of alternate sources of food, particularly protein supplements. Some of the most interesting of these are as follows:

1. Single-Cell Protein (SCP). Certain strains of single-cell bacteria (yeasts) will grow on petroleum oils, sewage sludge, and other media to produce an edible form of food known as *single-cell protein.* Active research has been carried out by the international oil companies, and experimental plants have been built to produce thousands of tons annually. To date, essentially all the material has been used for animal-feed protein supplements. Attempts to produce SCP for human consumption have been less successful because of consumer resistance to the product; severe gastrointestinal upsets often occur, and there is always the potential medical danger of consuming a product from petroleum. It is unlikely that SCP will become an important source of human food in the foreseeable future, although its increased availability for animal feed could release large amounts of other protein food such as fish meal for human consumption.

It is just possible, however, that a recent development in the United States may change the SCP picture. The federal government is spending large quantities of money to develop synthetic fuels by converting wood cellulose to glucose. Several processes under consideration involve the use of enzymes and yeasts; if these prove to be commercially feasible, very large amounts of SCP may become available as an inexpensive by-product.

2. Protein from Algae. It has been found that certain algae contain appreciable amounts of protein, and experiments have been carried out to produce a powdered protein supplement from specially grown algae. It is also possible to grow algae in sewage sludge. Again, there would undoubtedly be considerable consumer resistance to such materials, and the only feasible use appears to be as animal feed (where there is less concern about the origin of the material).

3. Leaf Protein. A potentially large source of protein exists in the leaves of plants such as alfalfa, sorghum, and water hyacinths. Much of this protein is of a high quality, and research is being carried out on methods to recover it for use as animal feed or as a protein supplement for human consumption. Because of the many practical problems involved in processing large amounts of material to recover the protein, it is unlikely that leaf protein will become an important source of food in the near future.

4. Genetic Modification of Food Crops. Considerable research is being carried out on the production of entirely new kinds of grains with improved properties. A new variety of corn has been developed that has twice the protein content of ordinary corn. High-protein strains have been developed of sorghum, an important food crop in Africa and Asia. Nutritionally speaking, these high-protein varieties are much better than those commonly grown today; unfortunately, the yields are often less, and there is also some question as to whether they are more susceptible to insect attack since they may prove to be an improved food for insects as well as humans.

An important goal of current research is to incorporate a nitrogen-fixing capacity in grains similar to that which exists in legumes. If this could be achieved, the growing crops would utilize nitrogen directly from the atmosphere and thereby save tremendous quantities of nitrogen fertilizer. Reseach has been carried out on inoculation of corn with special nitrogen-fixing bacteria, and it is hoped that microorganisms naturally present in soil can be given nitrogen-fixing capability by means of *recombinant DNA* (deoxyribonucleic acid) techniques—that is, by altering the nature of the genes that transmit the hereditary characteristics of a species.

5. New Sources of Conventional Food. Probably the most practical way to increase the world food supply is to develop new food crops. For example, soybeans have grown rapidly in importance during recent years. The soybean has a high protein content combined with the advantage of being a legume and thus not requiring nitrogen fertilizer. Other foods of growing nutritional importance include peanuts, cottonseed oil, and safflower oil. These have the advantage of growing satisfactorily on marginal land in many instances, thereby increasing the total acreage under cultivation; in addition, they can be rotated with more conventional crops to improve soil fertility.

FERTILIZERS

Organic gardening and natural foods have received a great deal of publicity in the United States during recent years. These involve the growing of food crops with only natural fertilizers such as leaves, hay, and manure—in other words, without the use of commercial fertilizers or pesticides. While this is fine for the small or medium-sized home garden, it has proven to be quite impractical for large-scale agriculture. The problem here is that the supply of natural fertilizers is far too small to make any significant contribution to commercial food crops; in addition, we know that to produce food some use of pesticides is almost always required today.

There is no question that the higher yields made possible through the green revolution are strongly dependent on the use of fertilizers; in fact, the increased use of chemical fertilizers accounts for much of the recent increase in world food production. Since 1950, world production of chemical fertilizers has doubled about every ten years. In the United States alone, production of synthetic ammonia (75 percent of which ends up in fertilizers) has been doubling every nine years. During this same period, world food production has increased by approximately 75 percent.

A balanced fertilizer contains three elements essential for plant growth: nitrogen, phosphorus, and potassium. In the United States, almost all fertilizer nitrogen comes from ammonia, which is usually produced by chemically reacting steam, air, and natural gas. Fertilizer phosphorus comes from calcium phosphate rock, which is treated with either sulfuric acid or phosphoric acid to make the phosphorus available for plant use. Fertilizer potassium comes from soluble salts that occur in the ground at depths of 1000 to 3000 feet and are brought to the surface by conventional or solution mining methods.

Three Major Problems

The world is now experiencing major problems in its use of fertilizers to increase food production. First, the law of diminishing returns places a limit on the amount of fertilizer that can be used to advantage. When a balanced fertilizer is applied to hybrid corn in the midwestern United States at the rate of 40 pounds to the acre, each pound of fertilizer increases the yield by 27 pounds of corn. When an additional 40 pounds of fertilizer per acre is applied, the increase in yield is only 14 pounds of corn. Additional 40-pound increments of fertilizer increase the yield by 9, 4, and then only 1 pound. An economic limit is reached where the cost of the added fertilizer becomes more than the value of the additional corn. A similar situation exists with other grains, such as wheat and rice. The United States, western Europe, and Japan are all now using fertilizers at about the maximum feasible rate, and it is very unlikely that there can be any substantial increase of crop yields in these areas. Other countries, such as India, Argentina, and possibly China, still use only small amounts of fertilizer; it is possible that food production would be appreciably increased in these areas by additional use of fertilizer.

A second problem facing the world is the cost and availability of fertilizer. While no shortage of phosphate rock or potassium is seen for the immediate future (at least in the United States), almost all of our fertilizer nitrogen comes either from natural gas or from petroleum distillates such as kerosene, both of which have experienced periodic shortages in recent years. In the United States, the cost of fertilizer nitrogen actually decreased substantially through 1971; however, the increased cost of energy (both petroleum and natural gas) since 1973 has caused a corresponding increase in the cost of fertilizer nitrogen. We can look forward to continued higher costs, periodic shortages, and growing problems of fertilizer supply for everyone, particularly the undeveloped nations.

The third problem today is the increasing pollution resulting from the widespread use of fertilizers since the end of World War II. Phosphorus leaching into rivers and lakes has increased the growth of algae. Nitrates originating in runoff from land treated with chemical fertilizer have appeared in the groundwater in many parts of the United States. These are poisonous to farm animals; even more serious, doctors are now aware that even several parts per million of nitrate in drinking water can cause cyanosis and death of infants.

It is also entirely possible that the cost and availability of energy may become the limiting factor in worldwide food production. Besides the direct use of petroleum and natural gas to produce fertilizer, large amounts of energy are required in the planting, cultivation, and harvesting of food crops, in processing foods, and in getting the final product to the consumer. According to Department of Commerce figures, the food industry is the fourth largest consumer of energy in the United States. About half of all trucks on the highway today are carrying food and related items. In addition, 60 percent of the nation's water supply goes to agriculture. Today, one American farm worker feeds fifty-six people. The entire food industry from farm to supermarket has become big business, and only a continuous supply of energy, fertilizer, insecticides, and other raw materials can keep this monster functioning.

PESTICIDES

In recent years, the use of pesticides has come under increasing attack by environmentalists and natural-food advocates, who consider every pesticide to be harmful. Although there is no question that pesticides have often been used in a very irresponsible way, under present conditions, large-scale production of food in the United States is impossible without them. Modern food production, which permits the United States to feed a substantial fraction of humanity, is based on the generous application of synthetic fertilizers, pesticides, and herbicides.

The first pesticides to be used were largely inorganic compounds, such as lead arsenate, copper sulfate, and cryolite. Shortly before World War II, natural organic pesticides, such as nicotine sulfate, pyrethrum, and rotenone, were discovered. This led to intensive worldwide research to determine the effectiveness of other organic chemicals, particularly those that could be produced synthetically from inexpensive and available raw materials. This culminated in the discovery of the powerful chlorinated synthetic organic pesticides such as DDT, chlordane, dieldrin, benzene hexachloride and the effective herbicides 2,4-D and 2,4,5-T.

Following the war, these synthetic organic chemicals were widely introduced as "miracle" pesticides which would permanently solve all our agricultural problems and many others as well. Production increased sharply, and by the 1960s over 100,000 tons per year of these chlorinated compounds were being used for agricultural purposes. DDT was everywhere, sprayed from the air on forests, orchards, vineyards, and vegetable crops, dusted on home gardens, and marketed in push-button sprays to kill household insects and in powder form to kill roaches. In the thirty years following World War II, over 1.5 million tons of DDT were produced in the United States alone.

As early as the 1950s, however, questions were being raised whether this wholesale use of DDT was a good idea. It was found that DDT might be too effective, that it often wiped out beneficial insects and thereby permitted the growth of parasitic species which were not normally a problem. It also turned out that DDT was concentrated strongly in going up the food chain: a trace amount of DDT in water would be concentrated to perhaps a tenth of a *part per million* (ppm) in plankton, 1 or 2 ppm in fish, and 10 to 20 ppm in fish-eating birds. Furthermore, it was found that even low levels of DDT in birds interfered with their metabolism of calcium, which resulted in their laying of eggs with shells so thin that they were crushed by the weight of the incubating parents. This actually caused a serious decline in the population of bird species such as falcons, eagles, ospreys, and pelicans. In 1962, Rachel Carson published her book, *Silent Spring,* which did much to publicize the problem of pesticides in the environment.

By the 1960s, another problem had appeared; strains of pests were developing that were resistant to the new pesticides. In 1964, the President's Science Advisory Committee reported, "By the end of the 1963 season, almost every major cotton pest species contained local populations that had developed resistance to one or more of the chlorinated hydrocarbons, organic phosphorus or carbamate insecticides, or mixtures of chlorinated hydrocarbons." This began a trend toward heavier applications and the use of increasingly powerful pesticides.

Also by the 1960s, it was gradually being recognized that DDT was not only an extremely persistent pesticide but was also being rapidly distributed throughout the entire world. Traces of DDT were found in fish and marine life taken from the middle of the ocean, in Alaskan seals, in Antarctica's fish and penguins. Most of the United States population had accumulated several parts per million of DDT in their body fat. Even the milk of nursing mothers would have been declared illegal and unfit for interstate commerce because of excessive concentrations of DDT at this time. According to a half-humorous newspaper account, certain cannibal tribes were no longer interested in white humans—DDT in their body fat was spoiling the good old-fashioned flavor!

Ideally, a pesticide should have a stability just long enough to perform its function and then disappear completely with no persistent residue. Table 4 shows that, in the case of chlorinated insecticides such as DDT, several years are required to remove any substantial fraction from soil, and there have been studies indicating that DDT may persist for as long as twenty years in some parts of the environment. Because of the many problems associated with the chlorinated insecticides, their use in the United States has now been banned except for emergency situations which require a special permit. The most important pesticides now in use are the organophosphorus compounds, such as malathion and parathion, which degrade very rapidly to leave essentially

Table 4. Time Required for 75 Percent of a Pesticide to Disappear from Soil (Assuming Normal Agricultural Conditions)

Pesticide	Type	Time for 75 Percent Removal
Chlordane	Chlorinated	5 years
DDT	Chlorinated	4 years
Dieldrin	Chlorinated	2 years
Benzene hexachloride	Chlorinated	2 years
Diazinon	Organophosphorus	12 weeks
Malathion	Organophosphorus	1 week
Parathion	Organophosphorus	1 week
CIPC	Carbamate	8 weeks
IPC	Carbamate	4 weeks
2,4-D	Herbicide	1 month
2,4,5-T	Herbicide	5 months

no residue. Unfortunately, they are extremely toxic to man and domestic animals, and there have been many cases of illness and some deaths associated with the use of these compounds.

Most growers today are moving toward *Integrated Pest Management* (IPM), which accepts the fact that it is impossible to completely wipe out a given pest. Control is achieved by using a pest's natural enemies, controlling pest breeding environments, or building pest resistance into crops and farm animals. Under IPM, chemical pesticides are used only when necessary and in amounts limited to controlling the pest population at a tolerable level consistent with acceptable (but not necessarily maximum) harvests.

Although the use of DDT has been essentially stopped in the United States, this is not true in the rest of the world where large quantities are still used to control insects. The present level of malaria control could not have been achieved without DDT; if its use in the tropics were stopped, it is entirely possible that there would be an additional one or two million deaths each year from malaria. It is not surprising that in certain countries, such as the Philippines, traveling exterminators continue to visit the villages regularly to spray DDT on the interior walls of the houses.

HERBICIDES AND RODENTICIDES

Food production in the United States is also strongly dependent on *herbicides,* which are used to clear land of brush and shrubs, to kill weeds (eliminating the need for cultivation), and to defoliate crops such as cotton to make harvesting easier for mechanical pickers. A recent development in agriculture is the use of *zero tillage,* in which a massive application of herbicides completely eliminates the need for plowing. Two substantial advantages are claimed: (1) Soil erosion by wind or water is greatly reduced, since the top layers are undisturbed and the plant roots tend to better anchor the soil. (2) Plowing, which can cause certain marginal types of soil to become badly compacted with a corresponding loss in fertility, is eliminated.

The most widely used herbicide is 2,4-D, a synthetic plant hormone that is very effective against broad-leaved perennials (including many of the common weeds). This is the notorious "Agent Orange" used in Vietnam to defoliate the forests. There appears to be considerable question whether or not 2,4-D is carcinogenic; animal tests have been reported negative, but the incidence of cancer among a number of Vietnam veterans has caused uncertainty about further use of this material. It would certainly be wise to use it with much caution.

Rodenticides are used to control food loss during storage, a particularly serious situation in warm climates. For example, it is estimated that rats outnumber human beings by about three to one in India—and that 10 percent of India's food is lost to rats. In warm climates, storage of food for a year may result in 50 percent loss. In Africa, birds eat a large fraction of the grain, and in some regions religious beliefs hinder effective food storage. Such losses obviously decrease the amount of food available to the ultimate consumer, and it is clear that even modest efforts to improve storage methods would be very effective in increasing the world food supply.

FOREIGN AID

Large-scale foreign aid by the United States began immediately following World War II with the very successful Marshall Plan. Although there were undoubtedly humanitarian feelings involved, the chief purpose of the Marshall Plan was to rebuild Western Europe as a bulwark against communism. With some $12 billion of American help, the reconstruction of Western Europe was achieved in a much shorter time than had been thought possible—largely, of course, through the bootstrap efforts of the Europeans themselves.

Despite this success in Western Europe, the United States government was still worried about the undeveloped nations of the rest of the world, which represented some two-thirds of the entire world population. China had already adopted a communist form of government, and it was feared that India, Indonesia, Burma, Latin America and Africa might be next. We needed their trade and raw materials to maintain our high standard of living; furthermore, the prospect of being one of very few democratic nations left on the face of the earth was not a comfortable one.

It was at this point that a minor employee of the State Department named Ben Hardy had an idea: Why not extend the Marshall Plan to the rest of the world? State Department officials felt this was too vague and premature, but Hardy passed it along to a young White House assistant who promptly turned it over to his boss, President Harry Truman. In his inaugural address on January 20, 1949, President Truman announced to the world that the United States was embarking on ". . . a bold, new program for making the benefits of our scientific advances and industrial progress available for the improvement and growth of undeveloped areas. . . ." All nations would work together through the United Nations. ". . . to help the free peoples of the world. . . achieve the decent, satisfying life that is the right of all people."

In the late 1950s, economist Walter W. Rostow proposed a new concept for foreign aid, the *takeoff point.* The economy of an undeveloped nation was similar to an airplane—all that was necessary to get it flying was to accelerate it to the point where the economy could take off from the ground, and then it would be on its own. No more foreign aid would be needed!

What has been the result of all this? There is no denying that the globalized Marshall Plan has been almost a complete failure. During the past thirty years, the United States has poured roughly $100 billion into genuine aid plus substantial and controversial amounts of military aid. With very few exceptions, the undeveloped nations have not achieved a takeoff point; the gap between the rich and poor nations of the world is virtually unchanged.

Probably the main reason foreign aid has been such an abysmal failure is our naive nineteenth-century evangelism which tries to "make everyone else just like us." It is one thing to rebuild postwar Europe; it is something else altogether to attempt to transform India into a modern industrial nation. A more basic approach must be taken in determining the needs of different cultures. We must recognize, for example, that not every nation in the world will function best as a democracy. For some nations, only very limited goals have any chance of success. In some instances, supplying good shovels may be far more effective than providing a tractor.

Foreign aid also demonstrates the same fundamental weakness as a domestic welfare program: once started, it tends to propagate itself on a continuously increasing scale. Although the American public in general is less than enthusiastic about the foreign aid program, industrial and political interests continue to keep it alive. Much foreign aid is admittedly an attempt to buy friends in sensitive areas of the world (Iran, Turkey, Israel, and Egypt are examples). At the same time, of course, our farmers are selling surplus grain; railroads and shipping lines, fertilizer companies, manufacturers of farm equipment, banks and automobile agencies are all reaping profits from the program. These interests, plus government agencies such as the Agency for International Development (AID), private philanthropies, many United States and foreign business organizations, and the United Nations itself have had thirty years of lucrative fallout from the foreign aid program. It is no coincidence that most of the pressure on Congress favoring continuance of foreign aid comes from organizations that have a vested interest in the program.

Perhaps the best illustration of bureaucratic thinking was presented by Miles Copeland in a December 21, 1973, letter to the *Times* of London. He wrote that all the five-year plans, seven-year plans, and such for Third World nations on file at that time in the offices of the Agency for International Development would, if implemented, require seven times the total available world supply of energy. When Copeland had presented these findings to friends in AID, "each was able to produce a convincing defense of the plans for his particular country but was oblivious of the extent to which his country's plans were in conflict with the total."

Today the United States often finds itself regarded pretty much as "Uncle Sap" by the rest of the world. The only real friends we have are those we would have had anyway, without a penny of foreign aid. On one hand, we are told we should drastically increase our foreign aid commitments; at the same time, however, we are openly accused of imperialism and neocolonialism. The fact is that during these thirty years of foreign aid programs, most of the undeveloped nations have looked on the Soviet Union as their friend in spite of the very small amount of foreign aid given by the Soviets.

China and India

It is interesting to once again compare the present situation in the world's two most populated nations, China and India. Fifty years ago, both countries were overpopulated, miserable, and facing a dismal future. In the 1930s, the communist revolution began in China with the basic philosophy of "tzu li keng sheng" (regeneration through our own efforts). During the years of civil war and of war with Japan its soldiers were encouraged to help with the planting and harvesting of food crops. Following the war years, contact between the West and mainland China was cut off entirely and, for a time, Soviet Russia contributed aid. In 1960, the Soviets canceled contracts and withdrew their technicians because of a conflict between Soviet and Chinese doctrines. Forced to go it alone, the Chinese under Mao Tse-tung, completed projects started by the Soviets, greatly expanded food production, and instituted effective programs for birth control.

Official figures indicate that China has been able to substantially lower its rate of population growth. Annual food production per capita has not changed significantly since 1955, but it has at least kept up with the growth of population. Diplomatic relations between mainland China and the United States have been restored. With over a billion inhabitants, China still faces tremendous population pressures; however, its leaders appear to be fully aware of the problems they face—and one can have cautious optimism about the future of China.

India, on the other hand, has been "fortunate" enough to have had access during its years of growing independence from the British to tremendous amounts of foreign aid from the United States and other Western nations. The population growth rate has remained at about 2 percent for the past thirty years, during which time its population has almost doubled. At its present rate of growth, India's population will reach a billion by the end of the century. Thanks to the green revolution, increased irrigation and the introduction of modern farming practices, food production has been substantially increased; however, the average inhabitant of India still has less than 2000 daily calories of food with protein substantially below the recommended value.

The failure of the monsoon rains of 1966 and 1967 is clear evidence of India's potential for disaster. The late 1970s saw a few years when surplus grain was produced, but these have been followed by marginal crops. Increased imports of energy and fertilizer will be needed for any further expansion of Indian agriculture—and this at a time when relations between the United States and Indian governments are at an all-time low. With the admitted failure of its birth control effort, as we saw in Chapter 2, it has become impossible to view the future prospects of India with any degree of optimism.

What is Triage?

In military medicine, the term *triage* has been used in those times when the flow of wounded soldiers fills the tents of the battlefield hospitals and it becomes impossible for the medical staff to care for all of them even in the most rudimentary way. Under these conditions, the wounded have been arbitrarily divided into three classes according to the severity of their wounds:

Class 1. The "can't be saved"—those wounded so seriously that they cannot survive regardless of any treatment given to them.

Class 2. The "walking wounded"—those who will survive without any treatment, even though some may be suffering intense pain.

Class 3. The "perhaps they'll make it"—those who may possibly be saved by immediate medical attention.

In the use of *triage,* then, all medical efforts are devoted to those in the third class, and the first two are given no treatment at all.

Triage obviously requires that someone "play god." In dividing the wounded, the number placed in Class 3 is always limited by the number of medical doctors available, and marginal cases may sometimes have to be placed in the other two classes. *Triage* places a terrible burden on those persons who must do the classifying; yet, while it seems cold-blooded and inhuman, it is a means of saving the maximum number of lives. Devoting medical attention to those in Class 1 or Class 2 would mean certain death for some soldiers who might otherwise have lived.

In 1967, William and Paul Paddock, in a book entitled *Famine 1975!,* shocked many people by suggesting that the concept of *triage* be applied to the distribution of United States surplus grain. According to their proposal, the hungry nations of the world would be divided into three classes:

Class 1. Nations with inadequate leadership and where population growth has already greatly passed the agricultural potential. To send them food is ". . . to throw sand in the ocean."

Class 2. Nations that have adequate agricultural resources and/or foreign exchange for the purchase of food from abroad. These require no food aid at all to survive.

Class 3. Nations having an imbalance between food and population, but where emphasis on agriculture and effective birth control offer the possibility of bringing these back into balance. To provide time for these nations to make the necessary changes, all food aid would be devoted to Class 3, with the first two receiving nothing at all—as in the case of military *triage.*

The concept of tying foreign aid to effective birth control practices in the recipient country has been frequently suggested. In 1960, William Vogt, in his book entitled *People,* wrote:

> The realities of this situation would seem not only to justify but to dictate the inclusion of birth control in our foreign aid programs. We have been warned that to press birth control on any country would be politically disastrous. That I find hard to believe. . . . In the long range, with few exceptions, it will be only those underdeveloped countries that have cut their birth rates that will escape misery and chaos.

In 1978, economist Kenneth Boulding wrote in his book *From Abundance to Scarcity:* "Unless grants from the rich to the poor countries can be tied to an effective method of population control, they are likely to be ineffective and may simply increase the ultimate sum of human misery and the size of the eventual disaster."

Of the total worldwide exports of grain in 1978, the United States supplied 45 percent of the wheat, 34 percent of the rice, and 70 percent of the corn. The only other nations with any substantial excess production of food were Canada, France, Australia, and Argentina. Since World War II, the United States has sent abroad tremendous quantities of food as outright gifts or long-term "loans"; aside from some tokenism, the attitude of the other nations has been pretty much "cash on the barrel-head!" While the United States is the only country in a position to lead the way to more effective worldwide food distribution, it is very unlikely that our policies will change in the near future.

The fact is that, because of the green revolution, there has been no need dire enough to require application of the principles of *triage* to our foreign aid. There is absolutely no question, however, that sooner or later world population will overtake world food production, and mass starvation will then become a way of life. When that day comes, the United States will have to decide such things as whether to feed India or Mexico, whether to ration food at home to supply more abroad—and which of our friends we shall allow to die.

Finally, as was pointed out by the Paddock brothers, we must all face the fact that, when a 10,000-ton freighter loaded with American wheat leaves port, some 200 tons of nitrogen, 50 tons of potassium, and 41 tons of phosphorus also leave our shores. In addition, the land needed to grow the wheat has permanently lost some of its topsoil and fertility. When these figures are multiplied by the thousands of freighter loads that have been shipped out during the past thirty years, it is clear that a very significant portion of our natural resources is involved.

Chapter Four

An Energy Primer

Man's use of energy goes back to the discovery of fire, which became essential for his survival in many parts of the world when the last great ice age forced him to adjust to colder climates and to foods that required cookery. The Neolithic revolution and resulting development of cities placed increasing pressure on fuel reserves, and the world's first "energy crisis" hastened the collapse of its first great civilizations. The valley of the Tigris and Euphrates Rivers, the Holy Land, Greece, North Africa, and Italy were all in turn gradually deforested and, in many cases, became permanently barren and unproductive land.

While there is no doubt that our world will always experience shortages of various kinds, in most instances we can reasonably expect to find acceptable substitutes. Energy is a different story: there is no substitute for it, and there is no way of creating it. To say that "something will turn up" merely dodges the issue—our only real option is to use our supply to the best advantage. We require energy for heat in cold climates, most of our food must be cooked, energy is required to produce clothing, transportation depends on energy; the list is endless, but the fact remains that energy in one form or another is essential for human life throughout most of the world today.

Figure 11. Is this abandoned service station at Lean, New York, a symbol for the future? (Wide World Photos, Inc.)

SOME DEFINITIONS

To begin, let's be sure we are speaking the same language. In daily life, we use expressions such as "having enough energy to get up in the morning," or "going to work," or even "things just didn't work out between John and Mary." To the physicist, *energy* and *work* have very specific meanings. Energy is defined as the capacity for doing work, whereas work is defined as a force moving through a distance. Since this means little to the average nontechnical reader, a few examples may be helpful.

If a force is exerted on an object and the object moves, work has been done and energy has been expended. (The object must actually move; a person leaning against the side of a building exerts a force, but there is no work involved because there is no motion.) The energy that an object possesses as a result of its motion is known as *kinetic* energy. Natural sources of kinetic energy include wind, waves, and water in motion (for example, over a waterfall). On the other hand, the water in a reservoir at high elevation possesses *potential* energy since it can be run through hydraulic turbines at a lower level to produce work.

Many different forms of energy can be used to create the forces needed to produce work. Fuels such as oil and coal possess *chemical* energy since they can be burned to produce steam that will expand to do the mechanical work of driving a turbine or steam engine. Similarly, fuels such as uranium possess *nuclear* energy since they can be used in a nuclear reactor to produce steam. An electrical current possesses *electrical* energy since it can be used to drive an electric motor to produce work. Objects at high temperatures possess *thermal* energy since their heat can be used directly to produce steam; examples of this include *solar* energy radiated by the sun and *geothermal* energy from the hot molten core of the earth.

In addition to the fact that all forms of energy can be used to produce work, we have learned much about ways to convert one form of energy to another. For example, coal can be burned to produce high-pressure steam that is run through a turbine to produce work; by connecting the turbine to an electrical generator, the kinetic energy of the turbine can be changed to electrical energy; the electrical energy, in turn, can be used to pump water from a low elevation to higher elevation to produce potential energy. The field of *thermodynamics* is concerned with the relationships among the various forms of energy, particularly the efficiency that can be achieved in converting one form to another.

Take the case of *heat,* a form of energy that a material possesses as a result of the random motion of its atoms and molecules. Heat is considered to be a lower form of energy; although such forms as electrical or kinetic energy can be converted 100 percent to heat energy, when we go in the opposite direction by changing heat energy to kinetic or electrical energy, we find that only about one-third of the heat energy can be utilized and that two-thirds is wasted. Practically all forms of energy are eventually transformed into heat. When we step on the brakes of a car, its kinetic energy is changed to heat through friction of the brake linings. A power plant generating electricity uses water from a lake to condense its exhaust steam, but not without raising the temperature of the lake. Since any production of heat represents loss of a more valuable form of energy, then, heat production is often kept to a minimum in energy systems, for example through the use of high voltages in electrical transmission lines or through the use of lubricants to reduce friction in mechanical systems.

RENEWABLE VERSUS NONRENEWABLE SOURCES OF ENERGY

It is necessary to clearly distinguish between our renewable sources of energy (which we can consider "energy income") and our nonrenewable sources ("energy capital"). Renewable sources will be available to us as long as human life exists on the earth, regardless of their rate of use. The most valuable of these are hydro, solar, wind, tidal, and biomass, and they represent only a small fraction of our total energy use today. Although our renewable sources will become increasingly valuable, we must recognize their limits and beware of a false feeling of security.

Nonrenewable sources of energy include the *fossil fuels:* coal, petroleum, and natural gas. These were created within the earth over periods of millions of years as a result of high pressures acting on organic deposits of vegetable and animal life. Since the industrial revolution, the fossil fuels have been used at a continually increasing rate—as if, in fact, there were no tomorrow. At our present rate of consumption, petroleum and natural gas will be gone in 100 years or less, and coal will be gone in about 600 years. By the time these fuels are no longer available, we will need to have developed other sources of energy to take their place. The need to survive on energy income alone, the renewable sources, would inevitably force us to drastically lower our standard of living or our population.

Nuclear fuels such as uranium and thorium represent another nonrenewable source of energy. Since its development for military use during World War II, nuclear energy has become an important source of electric power generation and, in 1979, accounted for 11 percent of the total electrical production of the United States. Certain states are already far more heavily dependent on nuclear power, such as Illinois (45 percent for the Chicago area in 1978) and Connecticut (over half nuclear).

Present nuclear power systems in this country involve the use of uranium 235 as fuel and are thus making available an additional energy source roughly equivalent to our remaining petroleum reserves. Although we are using these reserves at a slower rate than petroleum, uranium 235 is obviously not an answer to our long-term energy needs. Research is currently being carried out both here and abroad, particularly on breeder-type reactors that are expected to increase by some hundred times the energy available in uranium. As we will see under "Electric Power," there are many serious problems to be solved in the utilization of nuclear energy; however, if we wish to maintain our present life style in this country, there is no question that it will be required to supply an increasing share of our energy needs.

THE FOSSIL FUELS

If a hole could be drilled deep enough at any place on earth, it would eventually run into some type of dense crystalline rock, such as granite, whose grains are packed together so tightly that the rock is impervious to water or other fluids. This system of dense rock is continuous over the entire surface of the earth and is known as the *basement* or *basement complex.* In many parts of the world, such as eastern Canada, Scandinavia, and much of Africa, the basement complex actually occurs at the surface. In other parts of the world, it has been covered with deposits of sedimentary rocks, such as sandstone, shale, or limestone. These deposits vary up to as much as 10 miles in thickness and often occupy basin-like depressions over the upper surface of the basement complex.

Sedimentary rocks were created from the sands and muds of the geologic past, buried and preserved over millions of years. In some cases, organic materials from plant and animal life were trapped in these sands and muds and slowly changed to coal, petroleum, and natural gas—the fossil fuels. *These sedimentary rocks are the only ones in which commercial quantities of the fossil fuels have ever been found or are ever expected to be found.* It must be pointed out that the world's chief sedimentary basins have been defined and that many of them have been thoroughly explored, so that it is very unlikely that there are any large deposits of fossil fuels in the world that have not yet been identified.

COAL

Some 100 to 300 million years ago, large areas of the world consisted of level swampy land or very shallow water with profuse plant growth. As the plants died, they fell into the water and formed layers often several feet in depth. Covered by water and sand, these layers of organic material were preserved from complete decay and gradually buried to greater depths by further depos-

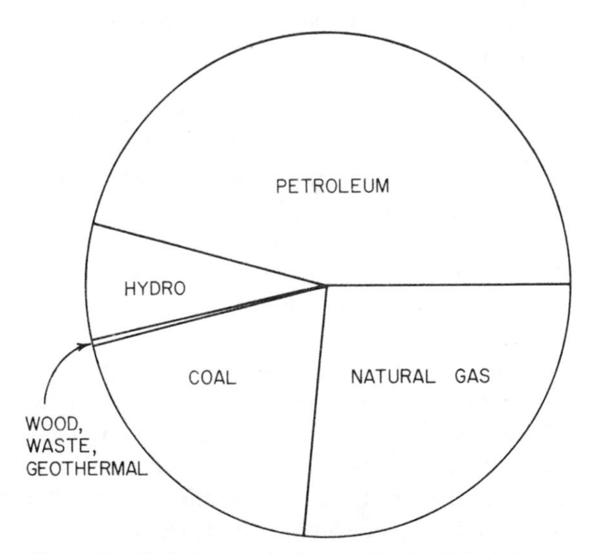

Figure 12. Today's sources of energy for the United States.

its of sand and mud and plants. Pressure and temperature increased with depth. The sand was gradually converted to sandstone and the mud to shale rock. The same combination of pressure and temperature also squeezed water and volatile matter out of the organic material, slowly converting it into the various grades or "ranks" of coal: peat, lignite, subbituminous, bituminous, and anthracite.

Peat is the first stage of coal formation. It is a black material consisting of partly decayed plants, reeds, and mosses growing in bogs. Because of its high moisture content, it must be dried before it can be utilized as a fairly low-grade fuel. Although peat occurs in many parts of the world, it finds only limited use as a fuel (in Ireland, for example) because of its low heating value. Peat occurs rather widely in the upper midwestern United States, but very small quantities are used for fuel.

Lignite, known in Europe as "brown coal," represents the second stage in coal formation. It is black or brown, crumbles easily, and appears to be made up of decayed woody-like material. Large quantities occur in Australia, South Africa, and northern Europe; in the United States, it is widely found in Montana and the Dakotas. Although lignite still contains substantial amounts of water and volatile matter, it has a sufficiently high heating value to be used directly as fuel. In South Africa and in parts of northern Europe, lignite serves as the chief source of energy.

Subbituminous and *bituminous* (sometimes called "soft coal") are the most abundant forms of coal in the United States. They have low moisture, moderate volatile matter, and a high heating value. Bituminous coals are always black and composed of layers that vary in appearance from a bright metallic shine to a dull and sooty surface. They fracture easily along planes or "cleats," which greatly facilitates the mining operation.

Anthracite (sometimes called "hard coal") occurs chiefly in Pennsylvania and represents the final process in the transformation from peat. It has very low moisture and volatile matter and consists mostly of carbon and ash. Anthracite normally has a heating value slightly lower than bituminous coal. It is very hard and shiny, has no banding, and breaks easily into small blocks. The chief use of anthracite in the past was as a domestic fuel since the low volatile matter minimized formation of soot. For many years, the only form of coal permitted in Chicago was anthracite.

From earliest times, it had been known that certain black rocks found on the beaches of Britain could be burned as fuel. These "sea coals" were simply small pieces eroded from exposed seams by wave action and were for many years only of local interest. At one time, in fact, both England and France are known to have banned the use of coal in their cities as a health hazard. By the thirteenth century, however, England's forests had been seriously depleted, and fuel was in short supply. The Admiralty was becoming concerned about the future availability of oak for the construction of warships, and royal consent was reluctantly given to mining coal in Newcastle. Coal was soon found to be an excellent substitute for wood in the smelting of metal and in other manufacturing processes of the time. With the development of the steam engine and the subsequent industrial revolution, Britain's coal production increased at a very rapid rate.

Although coal had been reported in America as early as 1634, it was not mined commercially until after the Revolutionary War. Public acceptance of coal was slow at first because "inexhaustible" forests were available for the taking, and the beginning of the nineteenth century found wood still accounting for more than 99 percent of our energy needs. Rapid growth of the railroads, beginning about 1830, provided greatly improved transportation, and soon the railroads became important carriers and consumers of coal.

By 1850, while coal still had little industrial use and over 90 percent of the blast furnaces were using charcoal, anthracite was becoming a valued domestic fuel in the big eastern cities of the United States. Rapid industrialization followed the Civil War and resulted in a doubling of our coal production about every ten years through the end of the century. In 1900, use of wood had just about come to an end, and development of other fuels such as oil and natural gas brought an end to the growth of coal.

Nevertheless, at the end of World War II, coal still accounted for more than two-fifths of our energy use. By the end of the 1950s, this had dropped to about one-fifth, production was static or even declining, and the price was about $5 a ton. In 1962, however, there was turnaround, and coal production began to increase steadily to meet the higher demand by electric utilities (which now account for about 75 percent of the consumption of coal). Since the Arab oil embargo of 1973, the price of coal has risen sharply and is now about $33 a ton (averaged delivered price to utilities), of which one factor is that about two-thirds of present consumption must be moved by rail at some $6 per ton.

The Future of Coal

In 1975, President Ford established a policy for doubling coal production in the United States to 1.2 billion tons a year by 1985. President Carter's 1979 energy program furthered this policy by forbidding the use of oil or gas in any new electrical generation plants and by encouraging the conversion of existing oil- and gas-fired units to coal. Production of coal has increased sharply in the last five years, but consumption has increased only modestly. This can be attributed to decreased growth in the use of electricity, to stringent air pollution regulations which limit coal use in some sections of the country, and to considerable reluctance on the part of the electrical utilities to convert (or, in some cases, reconvert) their oil- and gas-fired facilities to coal.

For many years, mining of coal was largely in underground mines located in the Appalachian region of Pennsylvania, Kentucky, Virginia, and West Virginia. These mines were close to the large Eastern markets, which benefited from low shipping costs. In recent years, however, there has been a gradual decline in the productivity of these mines because of stricter safety requirements. Coal from this area has become more expensive, and it is unlikely that this source will grow substantially in the future.

A second way to mine coal, of course, is surface or *strip* mining. This can be used when coal is present at depths of not over about 100 feet. The *overburden* (that is, the soil and rock deposited above the coal layers over the years) is removed to a "spoil pile" and the coal taken from the ground by large scoops —in some cases using explosives to first break up the coal. Surface mining now accounts for about 60 percent of coal production in the United States; it is considerably less expensive and much less hazardous, and a worker can produce three times as much coal in 1 hour from strip mining as in 1 hour of underground mining. Future increases in coal production will almost certainly come from surface mines.

Increasingly stringent air pollution regulations have created several problems that tend to discourage an increase in the use of coal in the United States. Solid materials (known as "fly ash") carried up the stack with exhaust gases —and in the past distributed over the countryside—are now generally removed by electrostatic precipitators. A far more serious problem today is the presence in the stack gases of sulfur dioxide. All forms of coal contain sulfur to a greater or lesser extent. When the coal is burned, the sulfur is converted to gaseous sulfur dioxide and goes up the stack with the other combustion gases. Sulfur dioxide can aggravate asthma and other respiratory diseases; in high concentrations, it also can kill vegetation. Of particular concern is that sulfur dioxide can react with water to form a mixture of sulfurous and sulfuric acid, producing an "acid rain" that may eventually change the natural acidity or alkalinity of the soil over very large areas located even hundreds of miles from the source of pollution. There is increasing evidence that vegetation in northern New England and eastern Canada is already being adversely affected by sulfur dioxide originating in the large coal-burning regions of our midwestern states.

Various suggestions have been considered to reduce or eliminate the problem of sulfur in coal. For example, there are widespread deposits of low-sulfur coal in our western states; if it were possible to use these more effectively, sulfur dioxide emissions could be substantially reduced. Because much of this low-sulfur coal occurs on government-owned land, both political and environmental considerations have delayed its development. These coal reserves can not only reduce pollution but also can be recovered by the less expensive surface mining, and it is clear that they are due for an important role in helping to meet the future energy needs of our country.

Research efforts have also been directed toward removing the sulfur from coal before it is burned. The use of certain organic solvents has been partially successful with some forms of coal, but much of the sulfur is bound so tightly in the coal molecules that it cannot be removed by ordinary extraction methods. Since these processes are quite costly, and since they remove only about half the sulfur present, it is unlikely that they will find widespread use.

Another approach to the sulfur dioxide problem is to remove it prior to discharging the combustion gases into the atmosphere. This can be accomplished by absorbing the sulfur dioxide gas in a solid material such as powdered limestone or in a liquid solution such as aqueous sodium carbonate. All such absorption processes are difficult and expensive because of the tremendous volume of gases produced by a large coal-fired plant. They increase the cost of electricity produced from coal by as much as 15 percent. In addition, any absorption process in itself produces large amounts of sulfur-containing sludge or solid residue that presents a formidable disposal problem. Because of these factors, electrical utilities have been very reluctant to adopt this method of sulfur dioxide control.

A less important but nevertheless significant problem in the use of coal as a fuel today is the substantial quantity of ash produced. Even without the use of any type of sulfur dioxide absorber, a large coal-fired power plant produces several hundred tons a day of solid residue. This alone represents a substantial landfill problem, particularly in highly populated areas such as the Eastern Seaboard—where many municipalities are having difficulty finding locations for the town dump.

The high shipping costs involved in getting coal from mine to market are a particular problem with Western coal, which is largely lignite and subbituminous and thus has a high moisture content and only about half the heating value of Eastern coal. Recently, for example, the municipal power system of San Antonio, Texas, complained to the Interstate Commerce Commission that the cost of moving Wyoming coal by rail to Texas was three times the mine price of the coal! Various transportation methods, such as the use of slurry pipelines, are being studied in an effort to reduce these extremely high transportation costs of Western coal, which are clearly inhibiting the development of important coal reserves.

At one time it was even proposed to build large electric power stations adjacent to the Western coal mines and then export the electricity, but most of the states concerned are now insisting that the coal must be exported—they have no intention of becoming "the dirty boiler room of the nation."

The fact is that, despite the efforts of our government to promote the use of coal, Rocky Mountain and Great Plains states now have a surplus capacity for the mining of coal that may reach some 120 million tons by the middle 1980s. This is the combined result of the decreased growth rate of the electric utility industry (which continues to be the main consumer of coal), the high shipping costs, and the environmental restrictions that limit new uses for coal. Much of the development thus far of Western coal has been by the oil companies, which view coal as a way to diversify and also a means to dispose of some of their current embarrassment of riches. They have not hesitated to spend the approximately $50 million required to open a new mine as a speculative venture without any sales commitments, either long or short-term.

Coal Conversion Processes

Many of the difficulties associated with the use of coal as a fuel can be eliminated by converting the coal to either a liquid or a gaseous form. The ash can then be removed by either filtration or settling, and the sulfur can be removed by conventional desulfurization processes, sometimes known as *sweetening,* used in the purification of petroleum and natural gas. The coal would thus be used as a raw material to produce synthetic liquid and/or gaseous fuels (*synfuels*). Coal conversion processes and our current progress are described in further detail under "Energy Alternatives."

Some of the conversion processes, of course, have been known for many years. More than 100 years ago, every large town had its own "gas plant," which produced gas for domestic cooking and for lighting needs. The gas was usually produced by the *water-gas reaction,* whereby red-hot coke was blown alternately with air and steam. Water gas had a fairly low heating value and was highly toxic because of its high carbon monoxide content. Although some parts of Europe still use a manufactured gas such as this, the United States has gone over almost entirely to other means of cooking, heating, and lighting.

Modern processes for coal conversion are all based on research carried out in Germany between the two world wars. Having very limited supplies of petroleum, Germany set out to achieve self-sufficiency in energy; during World War II, their synthetic processes were able to provide a large share of the fuels required for military purposes. While none of these are economically feasible in the United States at present, the rapidly increasing cost of petroleum and other fuels makes it appear inevitable that sooner or later we shall turn to these synthetic processes to replace the gasoline, fuel oil, and natural gas that now come from petroleum.

Coal is by far the most abundant fuel that we have today; in fact, the United States possesses about one-quarter of the entire known world supply of coal. Several years ago, the Bureau of Mines estimated that we have over 400 billion tons of recoverable reserves, enough to keep us going some 600 years at our present rate of consumption. Yearly production is slowly increasing and is now at an annual rate of 700 million tons. There is absolutely no question that we shall have to resort to coal more and more as our supplies of petroleum gradually run out.

PETROLEUM

To have any real understanding of the energy problems facing us, we must first have an understanding of the origin and refining of petroleum. The word *petroleum* actually comes from the Latin *petra* (rock) and *oleum* (oil); as with all fossil fuels, it is found only in sedimentary rocks. Derived from plant and marine animal organisms buried and preserved over millions of years under the gradual accumulation of sediments, oil and natural gas were slowly squeezed out of the source mud and settled into reservoirs of porous rock, usually sandstone, limestone, or dolomite. These rocks have a pore volume of about 20 percent, and the pore space forms a three-dimensional network which is normally filled with water; in some locations, however, oil and natural gas have displaced the water in certain layers of the sedimentary deposit. These local concentrations of oil and gas in the sedimentary rocks are the source of all commercial production of petroleum.

When a well is drilled down to a petroleum layer, higher underground pressure forces the petroleum to the surface, sometimes almost explosively. When the pressure is released, lighter constituents separate out as a gas that is chiefly methane but also contains smaller amounts of other light hydrocarbons. Often in the past this natural gas was simply burned ("flared") to get rid of it at the well's surface. Today, of course, it finds a ready market as an important fuel and will be discussed at length later in this chapter. The mixture of hydrocarbons that remains as a liquid after passing through the surface separating facilities is known as *crude oil,* the basic raw material for the manufacture of most of our petroleum products.

Some wells produce only natural gas, but regardless of its source, natural gas also contains appreciable quantities of many of the heavier hydrocarbons that can be used to produce gasoline and other valuable products. Since about 1920, it has become customary to extract these heavier components from natural gas to produce the *natural-gas liquids.* The sum of the two liquid phases, crude oil and natural-gas liquids, is often combined statistically and classed as the *petroleum liquids.* This practice has frequently led to confusion in specifying production or reserves of petroleum, since the inclusion of the natural-gas liquids actually increases the crude oil figure by about 17 percent.

The history of petroleum goes back several thousand years, and its commercial development goes back well over a hundred years. Asphalt was known to the Babylonians, the Chinese had natural gas wells hundreds of years ago, and petroleum from Burma was exported to England as early as 1700. In the United States, seepage of "rock oil" had been noted in many areas, and oil wells began to appear around 1830. These first oil wells were purely accidental —someone drilling for water or brine would occasionally find petroleum or natural gas instead.

In 1859, however, the famous Drake well was drilled at Titusville, Pennsylvania, and for some reason aroused the enthusiasm of speculators. Thousands flocked to Pennsylvania, and fortunes were made and lost. Petroleum production increased rapidly, doubling about every nine years from 1875 to 1929. For some seventy years, petroleum and natural gas have accounted for practically all the growing energy needs of the United States, particularly the ever increasing use of gasoline, fuel oil, and diesel fuels. For example, in 1900, petroleum and natural gas supplied about 10 percent of the total energy needs of the United States. In 1940, just prior to World War II, this had increased to about one-half and by the late 1970s reached about 80 percent. These figures alone indicate clearly why we have an energy crisis today.

Occurrence

Known reserves of petroleum are concentrated in a small portion of the world's sedimentary basins in two arcs, one in each hemisphere. In the Eastern Hemisphere, petroleum occurs in an arc traversing North Africa, the Middle East, and the central Soviet Union. In the Western Hemisphere, the arc traverses Venezuela, the western portion of the Gulf of Mexico, the central United States, and northern Alaska. These two arcs account for over 80 percent of the world's known petroleum reserves.

These reserves of petroleum are further concentrated in a relatively small number of very large oil fields. Some thirty-seven *super*-giant oil fields contain about half of the known reserves, and some 300 giant oil fields contain an additional 30 percent of the reserves. The remaining reserves, which constitute 20 percent or less of the world's supply, are located in over a thousand widely distributed smaller-sized fields.

The critical problem facing the world today is that only the discovery of super-giant and giant oil fields can substantially increase the amount of oil available to us. The world's sedimentary basins are well-known; in view of the efficient and extensive petroleum exploration that has already taken place throughout the world, it is very unlikely that a significant number of such fields exist. Perhaps the situation has been best summarized by Shell Oil Company president Harry Bridges when he said, "There is not a hope in hell that discoveries anywhere in the world will replace the Middle East."

There are seventy major sedimentary basins in the United States, and all have been probed, some extensively. Petroleum production in the United States was begun in the East (Pennsylvania, New York, and West Virginia) but rapidly spread to the West and Southwest. At the time of World War I, Oklahoma and California were supplying over half of our production. The giant Texas fields were discovered in the 1920s, and by World War II Texas was supplying 44 percent and California 19 percent of our petroleum. By 1978, the Texas contribution was down to 34 percent and California's was down to 11 percent. Production in all the important petroleum-producing states, including Oklahoma, Louisiana, Texas, Wyoming, New Mexico, and California was decreasing, and only the fact that new production in Alaska was now contributing 14 percent of our petroleum prevented a substantial decline in total United States production. For the immediate future, we shall have to work very hard just to maintain our present rate of production of petroleum; by the start of the next century, we will either have developed alternate sources of liquid fuels or else we will have to do with considerably less.

Proved Reserves and Ultimate Reserves

An important consideration in the operation of an oil well is the control of pressure and production during the life of the well. Pressure will be highest, of course, when the well is first brought into service. To insure maximum ultimate production from a well, the rate of production must be controlled so that the pressure does not decrease too rapidly with time. If the oil is taken too fast at first, sufficient time will not be available for the underground oil to find its way to the well, and the eventual yield will be smaller.

As the well approaches its end of life, *secondary recovery* may be used to increase the yield. This involves high pressure injection of air, gas, or water into the existing well to produce hydraulic fracturing of the rock, thereby opening up new channels for the oil. In addition, the injected fluid displaces oil in the reservoir rock and drives it to other wells in the oil field.

At any given time, some oil wells will be just coming into production, others will have been producing for some time, and still others will be ending their productive life. The total quantity of oil that geological and engineering data indicate with reasonable certainty to be recoverable in future years from known reservoirs is called the *proved reserves,* or sometimes just the *reserves.* This includes the portion of an oil field already drilled plus immediately adjoining portions not yet drilled but reasonably judged to be economically productive. Oil that can be recovered by techniques such as fluid injection is also included under reserves. The proved reserve figures are based on existing economic and operating conditions and are thus subject to modification as new techniques are developed or as the price of oil rises.

Of even more significance in the United States today is the question of *ultimate reserves* (that is, the total quantity of all oil that has been produced to date *plus* current proved reserves *plus* any reserves that may be discovered in the future). Although there is uncertainty as to the magnitude of the reserves yet to be discovered (particularly offshore and Alaskan reserves), we must remember that the geology of the United States is very well-known and that most potential oil-bearing formations have already been investigated. It is quite unlikely, as was pointed out earlier, that any large undiscovered oil fields exist, and a reasonable estimate of future discoveries indicates that ultimate United States reserves equal approximately 220 billion barrels. It is particularly important for us to understand the meaning of this term, for the fact is that approximately one-half of this oil has already been used up!

Enhanced Recovery

The total amount of oil present in a reservoir is often referred to as the *original-oil-in-place.* Using conventional methods of recovery, only about one-third of this original oil can be brought to the surface, leaving two-thirds in the ground when the oil field is depleted. Because of this very low recovery rate, research is continually being carried out to find ways of getting more of this oil out of the ground.

All methods of *enhanced recovery* are far more expensive than conventional methods, but they do offer us a possibility for getting a larger percentage of the oil out of the ground and are discussed further under "Energy Alternatives." While not normally considered in arriving at the proved reserve figures, it has been estimated that enhanced recovery could increase United States oil production by a million barrels a day by 1990—if the price of the recovered oil should become high enough to justify the incredibly high cost of recovery.

Refining Crude Oil

Crude oils vary considerably in appearance: some are black, some are light amber, some are thick as molasses, and some are fairly fluid. Regardless of appearance and origin, all crude oils consist mainly of hydrocarbons and have a composition of about 85 percent carbon, 13 percent hydrogen, and smaller quantities of organic sulfur, nitrogen, oxygen, and ash. Thousands of different hydrocarbons are present in crude oil, but they can be divided into the following three types: (a) *paraffinic* hydrocarbons, which range from methane (natural gas) to heavy paraffin wax and include thousands of straight and branched chains of carbon atoms; (b) *cycloparaffins,* or *naphthenes,* which are those hydrocarbons based on saturated rings, such as cyclopentane and cyclohexane; and (c) *aromatic* hydrocarbons, which are based on the six-carbon-atom benzene ring. The aromatics are particularly valuable in the production of gasoline because of their high octane number; however, they are very toxic materials, and benzene has been identified as a dangerous carcinogen. Some familiarity with these components of crude oil is necessary for a basic understanding of our current problems.

The refining of crude oil involves two successive operations. First, the original crude is distilled to separate it into a series of products (sometimes referred to as *fractions*) ranging from light gases to heavy tars. (See Figure 13.) This primary distillation produces some fractions that can be used directly for the production of salable products such as gasoline and fuel oil; because of the large amount of heavy constituents, however, most of the original crude ends up as heavy oils which have little market demand.

The second major operation of a refinery involves conversion of these unwanted materials (light gases and heavy oils) to marketable products by means of chemical refining processes such as *catalytic cracking.* In the United States, about 45 percent of the crude oil ends up as gasoline, about one-third of it is converted to various grades of fuel oil, and smaller amounts become kerosene, jet fuel, lubricants, wax, coke, and asphalt (see Figure 14). In the rest of the world, where there is less demand for gasoline, a greater share of the crude is converted to the fuel oils. This distribution also varies to some extent with the seasons; production of heating oil is increased to accommodate the fall and winter demand, and production of gasoline is increased to accommodate summer driving needs.

Usually, the first step in the refining process is to wash the crude oil with water to remove salt and other water-soluble impurities that might cause

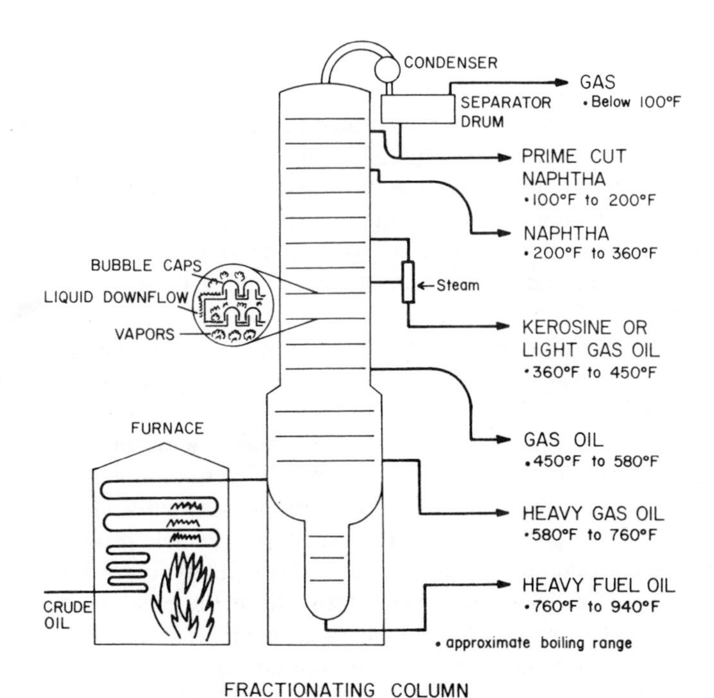

Figure 13. Primary distillation of petroleum.

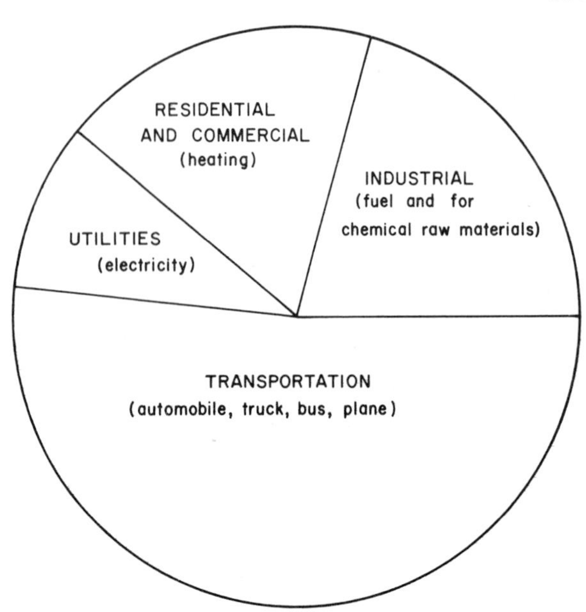

Figure 14. End uses of petroleum.

corrosion. The treated crude is then heated to remove methane and other light gases. (In the petroleum industry, the liquid fuel for an automobile is known only as *gasoline;* the word *gas* refers to volatile materials such as natural gas or bottled gas for cooking and heating.) Next, it is sent to a large atmospheric distillation column that separates a gas fraction at the top, a heavy residual oil at the bottom, and various side-streams ranging from light naphthas to heavy "gas oils." Each of these fractions is then subject to further refining. For example, the naphthas may be treated to remove sulfur and gum-forming materials and then go into gasoline blending stock. The heating oils are treated similarly, and they will end up as jet, diesel, and fuel oils such as residential heating oil. The heavy "gas oils" go to the catalylic cracker where they are broken down into lighter molecules and redistilled to produce more gasoline and fuel oil. By this complicated series of processes, roughly 90 percent of the original crude oil can be converted to readily marketable products.

Many of the refining operations, particularly catalytic cracking, produce large quantities of light gases, which result from the breaking down of the heavier molecules present. These light gases, combined with the gas fraction from the atmospheric distillation tower, are known as the refinery *off-gases.* At one time they were simply burned as fuel, but today they serve as a valuable by-product for the manufacture of the synthetic organic chemicals and polymers that have become so important in our daily lives.

THE PRODUCTION OF GASOLINE

The most important refinery product in the United States today is gasoline, and during much of the year refining operations are adjusted to maximize gasoline production. Certain requirements are placed on any fuel sold as gasoline: (1) It must have the proper *volatility,* which will depend on the place where it is sold and the time of year; for example, the oil companies appreciate cold winters for they must then use larger amounts of inexpensive butane in their gasoline to provide quick starts. (2) It must be *noncorrosive* to any parts of the fuel system; for example, sulfur must be removed to prevent the formation of sulfuric acid when the fuel is burned. (3) It must be *stabilized* to prevent the formation of gums or sludges that could block the carburetor. (4) It must have the proper *octane* number.

By far the most important of these requirements is the proper octane number. When gasoline is burned in an internal combustion engine, the air-fuel mixture must burn slowly to provide steady power to the piston during the power stroke. Under some conditions, the air-fuel mixture may start to burn and then suddenly explode with a knock or "ping" and a sudden loss of power. The tendency to knock depends on the type of engine, its compression ratio, and the particular chemical components of the fuel. The octane number is a standard laboratory measure of the ability of a fuel to resist knock when it is burned in a single-cylinder four-stroke engine of standard design. The higher the octane number, the less the tendency to knock.

Early in the history of the automobile, it was found that the molecules of branched-chain hydrocarbons and the aromatics such as benzene burn more easily and are thus most resistant to knocking, while the straight-chain hydrocarbons tend to knock. Two reference fuels are used in determining the octane number of a given gasoline: the straight-chain hydrocarbon normal-heptane is arbitrarily assigned an octane number of 0, and the branched-chain hydrocarbon isooctane is arbitrarily assigned an octane number of 100. If a particular gasoline knocks under the same conditions as a mixture of 10 percent normal-heptane and 90 percent isooctane, it is given an octane number of 90. The situation is unfortunately complicated by the fact that two standard sets of engine operating conditions are used in testing. The Motor (M) Method was originally used to grade all fuels. Development of higher octane fuels necessitated a different approach, which became known as the Research (R)

Method. Since this has sometimes presented problems in comparing different grades of gasoline, the octane value now posted at service stations is usually an average of the two, that is (R+M)/2.

One of the final and most valuable steps in the production of gasoline is *catalytic reforming* (not to be confused with catalytic cracking), where gasoline is passed over a platinum catalyst at a high temperature and pressure in the presence of hydrogen. Cycloparaffins are converted to aromatics, and straight chains are converted to branched chains, both leading to an improvement in octane number. Catalytic reforming also aids in the removal of sulfur and gum-forming materials from the gasoline.

Another method commonly used in the past to improve the octane number of gasoline has been the direct addition of tetraethyl lead (Ethyl fluid). This is the easiest and least expensive way to raise the octane number and for many years was used for all gasolines sold at service stations. Addition of lead directly raised the octane number by several points, thereby permitting a larger amount of the crude oil to be converted to gasoline at lower cost.

Unleaded Gasoline and the "Gas" Lines

In recent years we have all become quite conscious of periodic shortages of gasoline, particularly unleaded gasoline, in the United States. A look at the causes for this will show that short-sighted government regulation has been largely responsible. It is interesting to speculate whether the public would have accepted this regulation in the first place if they had been alerted to the great expense involved and the effect on the availability of gasoline for their cars.

In response to growing concern about air pollution, the Environmental Protection Agency (EPA) several years ago decreed new rules to govern the maximum permissible emissions from automobile exhaust systems. To meet the new standards, American car manufacturers placed catalytic converters on their new models—which was about as intelligent as taking aspirin for a broken leg! Any chemical engineer knows that catalytic processes are extremely tricky and are effective only with close control of temperature and operating conditions. Catalysts are easily deactivated or "poisoned" by overheating or by contamination from even small traces of impurities. How could anyone believe a catalyst would remain effective in an automobile using a variety of fuels and subject to periodic heating and cooling?

The most effective way to reduce emissions would, of course, have been to redesign the engine for more efficient burning of fuel. Not surprisingly, the automobile industry remained wedded to their powerful engines and saw the catalytic converter as the only acceptable solution to their problem—with the public left to face the consequences and pick up the tab.

Since lead is an excellent "poison" for the platinum catalysts that were to be used for emission control, it then became necessary to require unleaded gasoline for all new cars. Refiners were suddenly faced with the problem of producing tremendous quantities of high-octane fuel without the use of lead. To do this, they had to modify refinery operations, going to higher temperatures and pressures in the catalytic reformers and introducing other processes that would produce high-octane blending stock. This demanded the investment of hundreds of millions of dollars, which has been reflected in the higher cost of unleaded gasoline (a sharp-eyed motorist may have noticed that unleaded gasoline often has a lower octane number than regular leaded gasoline, in spite of its higher cost).

The less visible and more serious consequence of these new refining operations is their lower efficiency in producing gasoline. For example, the new higher pressures and temperatures cause undesirable side reactions which decrease the amount of gasoline produced by about 5 percent; in other words, the quantity of gasoline that can be produced from a barrel of petroleum is considerably reduced when a given octane number must be achieved without the use of lead.

A third consequence, at least in the case of the catalytic converters of the 1970s, has been an adverse effect on engine efficiency. The bed of catalyst raises the back pressure on the engine, sometimes resulting in a loss of efficiency. In addition, tuning the engine to meet the emission requirements often leads to a reduction of efficiency. Tuning must now be done to meet minimum emissions standards, which is inconsistent with maximum power or maximum gasoline mileage. In almost all engines today, the use of a catalytic converter is reducing to some degree the miles per gallon which can be expected.

The requirement to meet EPA emissions standards has necessitated the spending of billions of dollars (some of which has enriched platinum mine owners in South Africa), has reduced the quantity of gasoline available, and has reduced mileage. It is significant to note that these restrictions were formulated before the 1973–74 energy crisis. It also should be noted that the United States is the only country in the world to have adopted such regulations, some of which are now being relaxed by the Reagan administration. We can all hope that the environmental benefits will somehow justify the tremendous cost and the even greater waste of our energy reserves.

Growth of the Petroleum Industry

The development of the petroleum industry in this country actually represents one of the most remarkable scientific and engineering achievements of our industrial society. From the time of the first discovery of petroleum in the United States through World War II, petroleum production in this country has represented some 60 or 70 percent (or more) of the entire world production.

Large petroleum reserves in Texas and the southwestern states seemed to assure a practically unlimited future. These reserves at first easily supplied the domestic demand, and for many years the United States was an important exporter of crude oil and refined products. Rapid advancement in refining operations led to greater efficiency and reduced costs and prices for refined products. During World War II, the United States produced practically all the aviation gasoline and other petroleum products required by the allied nations.

With the end of World War II, the United States found itself the only major nation with unscathed cities and an industrial establishment untouched by war. Psychologically, this time also represented the end of the lean years which began with the depression of 1929 and continued through the "sacrifices" of the war years. The petroleum industry was ready for takeoff and anxious to do its part.

Much knowledge and experience had been gained from the operation of ocean-going tankers and pipelines as well as in the use of petroleum for products such as gasoline, fuel oil, nylon, and synthetic rubber. Oilmen now recognized that petroleum was a very versatile raw material, and they were quite prepared to supply oil and natural gas for the heating of buildings, diesel fuels to operate the railroad and trucking industries, and petrochemicals to produce plastics, synthetic fibers, medicines, fertilizers, and thousands of new chemical products based on petroleum and natural gas.

Two other factors had a strong influence on the growth of the petroleum industry following the end of the war. In 1945, coal still accounted for about half the total United States consumption of energy, and the coal miners were under the leadership of the colorful and militant John L. Lewis. In 1946, with the end of the wartime labor truce, Lewis led his United Mine Workers into a prolonged fifty-nine-day strike which produced much bitterness on all sides and substantially higher costs for coal. (As the newspapers of the time reported, "John L." was the best salesman the oil companies ever had.)

The second factor leading to postwar growth of the petroleum industry was the tremendous glut of oil that existed at the time. Drilling activity was at a very high level, more than doubling from 1945 to 1955, and the United States' proven reserves increased continuously until the 1960s, when they leveled off. Our oil companies suddenly became mammoth international corporations, with teams of technicians being sent throughout the world to explore for oil; "strikes" in South America and in the Middle East at that time turned up some of the world's giant oil fields. Rights to these were obtained under very favorable long-term concessionary leases, and the United States oil companies assumed that this was "their" oil, to be tapped indefinitely.

Between 1945 and 1970, the actual cost of producing Middle East oil was only about 10 cents a barrel; royalty payments to the local governments were $1 a barrel or less; with the cost to international oil companies under $2 a barrel, they could now ship it to the United States, and undersell domestic oil. This, of course, horrified members of Congress from our oil-producing states. Restrictive legislature was passed limiting oil imports as a way to protect the United States oil producers. Furthermore, at these low prices, oil had become competitive economically with coal, and because of their greater convenience oil and natural gas replaced much of the market previously held by coal.

Between 1950 and 1960, the number of motor vehicles in the United States increased by 50 percent, followed by a second 50 percent increase between 1960 and 1970. Between 1963 and 1973, the standard American automobile increased in weight by more than one-third of a ton, and gasoline consumption rose accordingly. In 1956, under President Eisenhower, Congress passed the Interstate Highway Act, which authorized a 42,500-mile network of limited-access roads to be funded 90 percent by the federal government and 10 percent by the states. The eventual cost of this was to be over $100 billion. In the early 1960s, the oil industry launched a national campaign to encourage more auto travel and other means to increase the consumption of petroleum. Energy experts were glibly forecasting that more oil would be consumed by the United States in the 1970s than had been used in the entire hundred years since Drake's first well in 1859. Yearly demand for oil in the United States was expected to double by 1985, and to double again by 2000.

During these giddy years, our country had essentially no national energy policy; it was strictly laissez faire as far as the oil companies were concerned. There were no complaints from the public, who seemed happy to buy the new gas-guzzling cars. To the politicians, "cheap energy" became synonymous with motherhood and apple pie, and influential congressmen from our oil-producing states made certain that nothing was done to rock the boat. The petroleum industry expressed eternal optimism, referring often to its "vast" new oil fields and the "fabulous" or "boundless" reserves yet to be discovered. Experts from the U.S. Geological Survey published official statements estimating petroleum reserves even beyond those of the oil industry—if a shortage of oil ever should develop, we could always import unlimited quantities of "our" cheap oil from the Middle East.

The Gathering Storm

Even during the postwar years of unlimited growth there were indications that the United States was headed for trouble. In 1948, imports of foreign oil exceeded exports—we had become a fixed net importer of oil. In 1951, President Harry Truman was concerned that future shortages of materials and energy might become a bottleneck to economic expansion and possibly jeopardize national security. He established the President's Materials Policy Commission with instructions to make an objective inquiry into all major aspects of the situation, and to make recommendations for a comprehensive policy on such materials. In June, 1952, a report was submitted to him giving forecasts of future needs for materials, energy, electricity, manpower, and technology for the years 1950 to 1975.

This report was published and made available to anyone by the Superintendent of Documents for $1.25 (paper cover). The Commission pointed out very clearly that the United States faced severe energy problems in the future. Estimates for future production and demand for crude oil were remarkably accurate: using a reasonable projection, they predicted that in 1975 the demand for crude oil would be five billion barrels a year (actual demand was slightly under six billion barrels), and that United States production would be slightly under three billion barrels a year (actual production was slightly over three billion barrels). It was predicted that domestic production would peak in 1967 and decrease thereafter. Peak production actually occurred in 1970 and has decreased since then. (See Figure 15 for chart of U.S. production and consumption of oil from 1960 to 1980.)

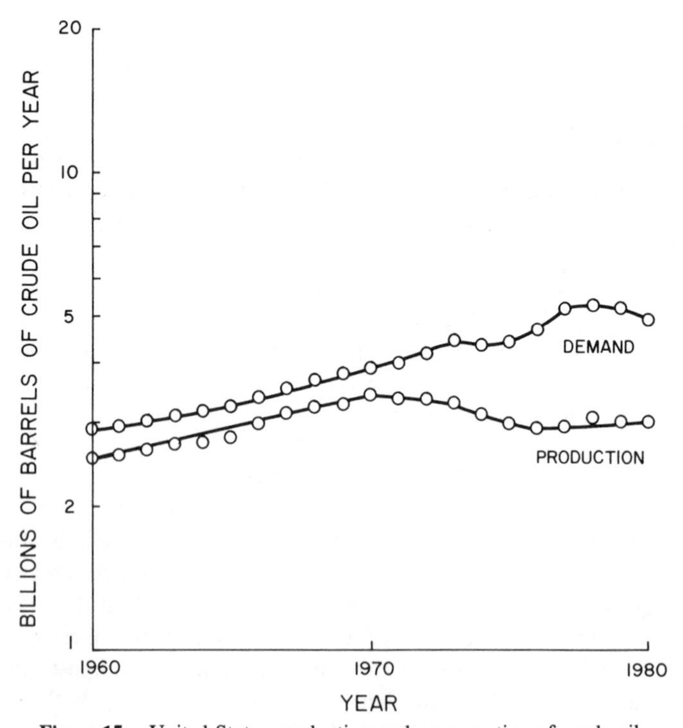

Figure 15. United States production and consumption of crude oil.

The Commission pointed out that the remaining reserves of oil and gas in the United States would be no match for the demands of the next twenty-five years.

> We will be forced to drill deeper, to drill more dry holes for every well brought in, and possibly to rely on new pools of smaller average size. United States demand may far outstrip domestic production, as it has started to do lately. . . . We will not suddenly run out of these highly important fuels, but we could well enter a long period of rising real costs.

The Commission recommended that the federal government encourage as soon as possible the construction of small-scale commercial plants to produce oil from shale. Of particular interest today was the Commission's comment on waste:

> The oil and automobile industries in particular carry a heavy obligation to lead the Nation toward a more efficient use of its liquid fuel supplies, and consumers themselves can contribute strongly if kept properly informed of conservation measures they can take.

In addition to the 1952 report by President Truman's Materials Policy Commission, the decade of the 1950s saw many other estimates of future crude oil production in the United States. There was general agreement that the ultimate reserves were in the range of 100–250 billion barrels, and that production would peak in the 1960s or 1970s and decrease thereafter. The most astute forecast was made in 1956 by Dr. M. King Hubbert, a geophysicist employed by Shell Oil. By applying the standard growth laws to the discovery and production of crude oil, Dr. Hubbert predicted that ultimate United States reserves were 200 billion barrels, and that production would peak in 1970 at three billion barrels a year and then decline. Production did, of course, peak in 1970 at 3.5 billion barrels, and ultimate United States reserves are now considered to be about 220 billion barrels (including Alaskan and offshore reserves).

Despite these warnings, the growth of the United States economy continued unchecked, based increasingly on higher oil imports. In 1950, our imports accounted for about one-sixth of our oil consumption. By the late 1950s, imports had increased to over one-quarter, by the late 1960s to more than one-third, and reached a maximum in 1977 when imports accounted for 45 percent of our oil consumption. United States proved reserves peaked in the 1960s and began a slow decline, which—except for a substantial jump in 1970 with the addition of the Alaska Prudhoe Bay Field—has continued through the present. In 1968, the State Department sent word to foreign governments that United States oil production would soon reach its limits, that in case of emergency the free world could no longer rely on any surplus United States capacity. By the early 1970s, Texas was producing at 100 percent capacity for the first time in forty years, and domestic oil production was on the decline.

Evidently the first clear recognition of the coming crisis was in July of 1971, when the National Petroleum Council, a government-industry study group which included most of the top oil executives, issued a report that predicted that, by 1985, the United States would be 57 percent dependent on foreign sources for supplies of petroleum liquids, and that these imports would constitute 25 percent of our total energy consumption. It is of interest to note that this report was a 180-degree switch from the uniformly optimistic forecasts made by many of these same experts during the 1960s. Secretary of the Interior Rogers C. B. Morton was so concerned about the report that he issued a statement pointing out the possibility of "a frightening energy scarcity" in the 1980s; at the same time, however, President Nixon assured us that the oil industry could supply the petroleum needed for the next generation.

Another factor that strongly affected the availability of crude oil was the rapid postwar industrialization of other nations of the world. In 1950, the United States consumed roughly two-thirds of the entire world production of crude oil. Rapid growth of Japan and western Europe then began to place additional demands on the supply. As our balance of trade worsened (largely due, of course, to our growing oil imports), it became obvious that Japanese and European currencies were considered "harder" than ours, and other countries have since become increasingly able to outbid us in the marketplace. By the 1970s, the United States was consuming only about 30 percent of the world supply.

Trouble in Paradise

It is generally agreed that our development of the oil reserves of the Middle East is an example of the worst type of foreign exploitation—and much of it with the direct encouragement and support of our government. As early as World War I (before the discovery of our large oil fields in Texas and Oklahoma), the United States became concerned about supplies of oil and adopted an unofficial policy of encouraging the acquisition of as many foreign sources of oil as possible. United States oil companies were naturally quite willing to cooperate in this profitable as well as patriotic undertaking. At the time, most of the oil rights outside of North America were owned by British Petroleum and Royal Dutch Shell. In 1920, Congress passed a law to prohibit any foreign-owned corporations from obtaining oil leases on United States public lands unless their own governments allowed United States corporations to explore for oil in the territories controlled by them. This opened the door for United States exploitation, and the race was on!

The first of the giant Middle East oil fields had been found in 1908 in Iran by British Petroleum. Starting in the 1920s, United States oil companies became more aggressive and soon had staked out a dominant position in the entire Middle East. Oil production began in Iraq in 1927 with Standard Oil of New Jersey (now Exxon) and Socony Vacuum (now Mobil) owning one-fourth of the Iraq Petroleum Company. A pipeline was constructed, and by 1934 Iraq had become a major exporter. In the early 1930s, the United States government persuaded the British to allow Standard Oil Company of California to explore for oil on Bahrain, a small island off the coast of Saudi Arabia that was a British protectorate; in 1932, Bahrain became the third Middle East producer. In 1934, the British permitted Gulf Oil to explore Kuwait, another British protectorate.

Discovery of oil on Bahrain in 1932 brought covetous eyes on the vast desert a few miles across the water in Saudi Arabia. In 1933, Standard Oil of California purchased oil exploration rights in Saudi Arabia for $300,000 in gold. Several years later, Texaco bought half the rights to Saudi Arabian oil, and the Arabian American Oil Company (Aramco) was formed. By 1939, Saudi Arabia was an important contributor, and today it is the dominant producer of the Middle East.

Following World War II, the United States recognized the value of Middle East oil, particularly Saudi Arabian oil. In 1945, the State-War-Navy Coordinating Committee (now the National Security Council) advocated construction of a United States military air field in Saudi Arabia and concluded:

> United States reserves are rapidly diminishing; that at the present rate of exploitation and consumption our reserves are adequate for perhaps 12 to 15 years. Thus the world oil center of gravity is shifting to the Middle East where American enterprise has been entrusted with the exploitation of one of the greatest oil fields. It is in our national interest to see that this vital resource remains in American hands, where it is most likely to be developed on a scale which will cause a considerable lessening of the drain upon the Western Hemisphere reserves. . . . Saudi Arabia possesses proved oil reserves that petroleum experts estimate at 5 billion barrels. . .

Today we know that the ultimate reserves of Saudi Arabia could easily be a hundred times this estimate.

On several occasions Aramco asked for (and got) United States government aid to protect its interests in Saudi Arabia. During World War II, President Roosevelt arranged for lend-lease assistance to Saudi Arabia to offset lower revenues from oil production and tourist pilgrimages to Mecca, both of which were suffering because of the war. In the early 1950s, Saudi Arabia was rapidly increasing its oil production and again asked for money. Aramco did not want to raise prices for their European customers, nor did they want to raise royalty payments to Saudi Arabia since this would reduce their profits. The United States National Security Council "resolved the issue" by arranging a mickey-mouse accounting procedure whereby Aramco received credit in the form of income–tax rebates—with United States taxpayers picking up the check. This effectively made foreign operations more profitable than domestic drilling and also ensured that higher world oil prices meant higher after-tax profits for Aramco.

In 1950, the State Department felt that the Middle East was "highly attractive and highly vulnerable" to communism. In 1951, Iran, under the leadership of Prime Minister Mohammed Mossadegh, took control of its oil industry by nationalizing British Petroleum's Iranian properties. The major oil companies responded by boycotting Iranian oil and were able to easily supply world markets by increasing oil production in other Middle East countries. In 1953, Mossadegh was overthrown by royalist troops and pro-Shah mobs (with the assistance of the United States Central Intelligence Agency), and the Shah was again in power. A consortium of oil companies then took over Iranian production—but, as far as the Iranians were concerned, the only difference was the fact that the Americans now took 40 percent of their oil. In 1955, five major United States companies controlled two-thirds of the Middle East oil, with the British controlling most of the remaining one-third.

World production of crude oil doubled between 1950 and 1960 and more than doubled again between 1960 and 1970. There was very rapid expansion of production in the Middle East and the Soviet Union. Africa, a negligible producer in 1950, expanded rapidly to 13 percent of world production. Oil production increased in North America and South America, but at a much lower rate. During this twenty year period, there was a glut of oil and prices were depressed, reaching a low of $1 to $1.20 a barrel in the Persian Gulf by the end of 1969.

During this period of time, revenues to the oil-producing countries were determined by the "posted price" which was established by the oil companies. In 1959, Exxon cut the posted price of Middle East oil—and cut it again in 1960. In accordance with the habit of those good old days, Exxon did not even discuss the move with the oil-exporting nations; it simply announced the price cuts. Of course, this led to considerable loss of revenue for the oil nations; they were furious, but there was nothing they could do at that time.

The world oil situation was starting to change by the early 1970s. Crude oil production in the United States, Africa, and South America had peaked and started a gradual decline. In the Middle East, crude production peaked at Bahrain Island, Kuwait, and Qatar—and began to level off in Iran and Iraq. Only Saudi Arabia was in a position to materially expand production to accommodate any future increased world demand for crude oil.

In 1960, Saudi Arabia, Iran, Iraq, Kuwait, and Venezuela had established the Organization of Petroleum Exporting Countries (OPEC) to unify their oil policies and promote their collective interests. The reaction of the oil companies was predictable: at the time they just pretended that OPEC did not exist. In May of 1970, however, a bulldozer accidentally broke a pipeline carrying oil from the Persian Gulf to Libya on the Mediterranean, where it was taken by tanker to Europe. This pipeline was especially critical because the Suez Canal was still closed as a result of the 1967 Arab-Israeli war. Libya seized this opportunity to squeeze the oil companies, and in 1971, led by the Shah of Iran, OPEC won a huge price increase of 50 cents per barrel.

Between 1970 and 1973, imports of crude oil by the United States more than doubled. In 1970, 18 percent of our imports came from the Arab countries; by 1973, this had increased to 34 percent. By the summer of 1973, world demand for oil was so great that even Saudi Arabia had to open all valves and go to maximum production.

THE ARABS USE THE "OIL WEAPON"

Since the end of World War II, United States policy in the Middle East had been based on the assumption that there were two types of nations there, those with oil and those without oil. Nations without oil were considered to be of no strategic importance. Nations with oil feared that the neighboring Soviet Union might some day try to take over their oil fields, and hence the Arab rulers of these nations would always in the last analysis be more anticommunist than anti-Zionist. Thus, as long as these rulers were on their thrones, the United States felt it could support Israel with impunity.

By the early 1970s, however, the United States "juggling act" was starting to come apart. The Arabs could see that we were becoming more dependent on their oil and that some of this oil was being used to supply arms for Israel. As the oil nations became more militant in their demands, the United States was gradually losing much of its influence in the Middle East. When King Idris of Libya was overthrown in 1969 by Muammar al-Qaddafi, a colonel in his army, Libya immediately demanded a major increase in the price of crude oil, threatening to cut off supplies to any oil company that did not meet its demands. This electrified the other oil nations, who followed through with similar price increases.

As early as 1970, when United States production of crude oil peaked, King Faisal of Saudi Arabia was warning American officials that this country would have to "alter its stand on Israel." In the spring of 1973, when our total imports of crude oil from the Middle East had risen to five times what they had been in 1970, King Faisal appeared on United States television, declaring, ". . .

America's complete support of Zionism against the Arabs makes it extremely difficult for us to continue to supply United States petroleum needs and even to maintain friendly relations with America." At about the same time, the Intelligence and Research Bureau of the United States Department of State forecast that there was roughly a fifty-fifty chance that war would break out in the Middle East and that, if it did, a disruption of Arab oil supplies could be expected—but, in the White House and on Capitol Hill, it was politics-as-usual.

When the Yom Kippur War started on October 6, 1973, Aramco warned the United States that any increased support of Israel would lead to a reduction in oil production by the Arab nations. President Nixon ignored the warning. The Arab oil producers were at first fairly restrained in their demands, but unexpected pressure from Kuwait (not one of the more active participants in the war) initiated an economic confrontation. After considerable discussion, a critical meeting was held on October 17; OPEC announced a 70 percent increase in its oil royalties, a 5 percent cutback in production, and threatened selected embargoes against any "unfriendly" nations. On October 18, President Nixon asked Congress for $2.2 billion in emergency military aid for Israel. From the American standpoint, this was little more than a financial transaction to cover military aid already provided to Israel; King Faisal chose to regard it as a personal betrayal of Saudi Arabia. On October 19, his Royal Cabinet announced that all oil exports to the United States were being suspended, and that King Faisal had called for a *jihad* (holy war) because of our increased aid to Israel. By October 21, there was a total Arab boycott of United States markets.

The amount of Arab crude oil coming to the United States at that time represented *only about 6 percent* of our total demand for petroleum products, and this could easily have been offset by a slight moderation of our wasteful energy habits. World supply of crude oil was reduced less than 10 percent by the Arab cutbacks, although certain nations of western Europe were in a worse position because of strong dependence on Arab oil. Widespread panic occurred. The United States saw its first "gas" lines and western Europe decreed automobileless Sundays. Refiners dependent on Arab oil were obliged to bid at auctions run by several OPEC countries. Iranian oil went for over $17 a barrel, and in Nigeria crude oil reached over $22 a barrel. When the ban was lifted on March 18, 1974, the price of crude oil had increased to some four times the price of a year earlier.

In the United States, a confused public looked for a scapegoat, variously blaming OPEC, greedy oil companies, the government, and even the local service stations. Many people simply lashed out at anything they felt might be responsible, and considerable latent anti-Semitism appeared. Neither the government nor the public seemed willing to put the blame where it really belonged: on the wasteful habits we had developed from a policy of cheap energy.

To combat the embargo, on November 7, 1973, President Nixon gave a televised talk to the people about the energy crisis which included the following recommendations: "Let us unite in committing the resources of this nation to a major new endeavor. . . we can appropriately call Project Independence . . . . Let us pledge that by 1980 under Project Independence we shall be able to meet America's energy needs from America's own energy resources."

Nixon's specific proposals to achieve Project Independence included increased use of coal by eliminating conversion from coal- to oil-fired plants, a speedup in the construction of nuclear power plants by reducing the time lag from ten to six years, a cutback in airline travel to conserve jet fuel, turning down thermostats in homes and offices to 68 degrees Fahrenheit, a national speed limit of 50 miles per hour, elimination of unnecessary lighting, staggered work hours, and encouragement of car pooling. He also asked the Congress to establish an Energy Research and Development Administration—the only one of his proposals to receive any real support in Washington at the time.

Project Independence was, of course, politically inspired and had no real prospect of achievement. Burdened with the problems of Watergate, President Nixon could expect no real support from Congress or the people. With the end of the embargo in March, our energy problems seemed to lessen, and there was a temporary glut of oil. Demand for OPEC oil actually decreased somewhat because of slower economic growth, conservation, and increased production from Alaska and the North Sea. Crude oil prices were periodically raised but not enough to keep up with inflation in the industrialized nations.

In retrospect, we might well ask whether the embargo was a direct consequence of the Yom Kippur War, or was it primarily an excuse on the part of the Arabs to achieve a substantial increase in the price of crude oil? Since the Arabs had repeatedly threatened an embargo over the years, it is entirely possible that it would have happened sooner or later. Also, although the Arab-Israeli war was ended only 4 days after the embargo was imposed, the embargo was continued for 5 months. In either case, as we noted in Chapter 1, the embargo was certainly the best thing that could have happened to the United States. We Americans are a peculiar people—we spend billions of dollars for reports, studies, workshops, evaluations, planning, and computerized projects, but we seldom pay any attention to what they tell us. As long as gasoline is available from the local service station regardless of its price, there is no shortage! The Arab embargo and the resulting "gas" lines served as a much-needed "kick in the pants" to make us realize the existence of an energy crisis and that something had to be done about it.

Problems in Iran

Since the fall of Mossadegh, the United States had looked on Iran as a stable ally, a military power that could maintain stability in the Persian Gulf. In the twenty years since 1953, this country had sold over $18 billion in arms to Iran. Iranian military and security forces had received extensive training from United States "advisers," and there had been considerable transfer of sophisticated weaponry. On December 31, 1977, President Carter visited Teheran and at a New Year's Eve party toasted the Shah with: "Iran under the great leadership of the Shah is an island of stability in one of the more troubled areas of the world. This is a great tribute to you, Your Majesty, and to your leadership, and to the respect, admiration, and love which your people give to you."

It soon became obvious, however, that the people's "love" for the Shah of Iran was nonexistent. During 1978, anti-Shah demonstrations increased in violence from month to month. The Shah placed Iran under military rule, and when the oil workers went out on strike, bayonets were used to force them back to work. Even this was unsuccessful; oil production became sporadic and eventually almost ceased. In January, 1979, the Shah and his family left Iran for a "vacation," and within a short time the Ayatollah Khomeini was supreme ruler.

Loss of Iran's 5 billion barrels a day of crude oil suddenly changed a worldwide oil "glut" into a shortage. Although several countries, particularly Saudi Arabia, increased production to take up some of the slack, the market continued very tight, and "spot" market sales jumped some $8 per barrel above the OPEC price. This was too much to resist, and soon the OPEC countries were adding surcharges to all their oil, even that covered by long-term contracts.

In December, 1978, OPEC had announced a price hike of 14.5 percent for 1979, much higher than anticipated. When Iranian oil came back on the market in March of 1979, other OPEC countries cut back production to maintain the tight market and thus continued to profit from spot prices. That same month, OPEC announced another price increase of several dollars per barrel, and by the close of 1980 crude oil was over $30 per barrel.

American relations with Iran plunged to an all-time low in November of 1979, when the American employees at the Teheran embassy were seized as "hostages." All Iranian assets in the United States were frozen, the United States stopped buying Iranian oil, and an embargo was placed on all shipments to Iran. When the Iran-Iraq war began at the end of 1980, there was a loss of several million barrels a day; once more a worldwide "glut" of oil suddenly became a shortage that inevitably triggered higher prices. This time, fortunately, our country was in a better position; our demand was down, and the oil companies had storage tanks overflowing with crude oil and refined products.

The Monday Morning Quarterback

There is a well-known saying in sports that "you can't replay the game" once it's over. It may be useful, however, to summarize the reasons for our present energy situation, not necessarily to attach blame, but to make sure we do not continue to repeat the mistakes of the past—and perhaps indicate possible solutions to the problems of today and tomorrow.

The first cruel fact that Americans must face is the depletion of our domestic reserves of crude oil. Over two million wells have been drilled in the United States, four times as many as in the rest of the noncommunist world. Two-thirds of the world's active oil rigs operate in the United States. There are twice as many oil wells in Kansas as in all of South America, three times as many in Arkansas as in all of Africa. Unless greatly improved recovery methods are developed, we must accept the statement that we have already used over half the total amount of petroleum in the United States—petroleum that required several hundred million years to create. During the 1970s, annual production of crude oil in this country averaged about three billion barrels, with newly discovered proved reserves averaging about 1.4 billion barrels. It is conservatively estimated that increased exploration during the 1980s may yield from 2.5 to 3 billion barrels of proved reserves annually. Accepting that figure, even though production will continue to decrease, we will continue to produce oil faster than we are discovering it. By the end of the century, our proved reserves may well be only a small percentage of what they are today.

From time to time, extravagant predictions are made of "vast undiscovered reserves" located offshore on our continental shelves. For example, in 1972, the U.S. Geological Survey estimated that the total domestic "resource base" for petroleum liquids was about 2900 billion barrels. In 1974, this Survey estimated that as much as 200 billion barrels of crude oil could be recovered from the continental shelf—only to issue a "revised estimate" a few weeks later reducing this to 100 billion barrels. At about the same time, a committee of petroleum geologists estimated that 300 billion barrels of recoverable oil could be found in the unexplored offshore areas of the United States, recommending that the federal government finance the drilling of test holes to confirm their estimates. To date, results have been far from spectacular: several wells have been drilled in the "promising" Baltimore Canyon offshore area without finding the predicted "vast" quantities of crude oil and natural gas.

Unfortunately, the situation in the rest of the world is little better than in the United States. Total world production of crude oil is now about twenty-two billion barrels per year, and it is very unlikely that this will increase significantly during the 1980s. Production actually deceased in 1980 because of the Iran-Iraq war. Oil production in the following important producing nations has peaked and begun to decline: United States, Canada, Brazil, Colombia, Venezuela, Rumania, Bahrain Island, Kuwait, Oman, Turkey, and Libya. The only nations that appear to be in a position to significantly increase production are Mexico, Saudi Arabia, China, and possibly Iran (when it can solve its internal problems). About 60 percent of the known world reserves are located in the Middle East. The United States has 5 percent of the world reserves and 5 percent of the world population, but we are using 30 percent of the world production of oil! Figure 16 charts the world production of oil from 1960 to 1980.

A particularly troublesome situation will develop in the Soviet Union during the 1980s. Soviet production of crude oil has just about reached its peak,

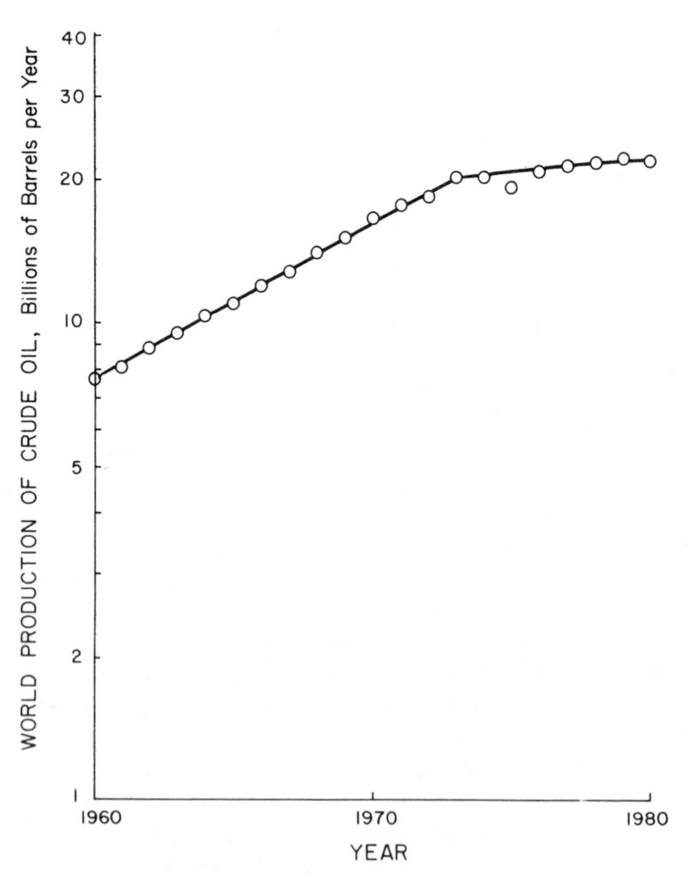

Figure 16. World production of crude oil.

during the 1980s the Soviets will probably change from an exporting nation to an importing nation. This will place even more serious pressures on the world supply. Furthermore, they will probably be looking to the Persian Gulf for their additional needs, so that prospects are very strong that an already dangerous situation in the Middle East will become far worse.

GOVERNMENT POLICY

There is no question that many of our problems today are directly attributable to the total lack of government policy on energy over the years. There was certainly plenty of warning of impending shortages, particularly in classified military reports, but much of it was more or less deliberately kept from the public. Our politicians seem to have one primary interest—getting themselves reelected—and they have consistently refused to take any of the hard steps needed to forestall our present situation since these would be unpopular and might rock the boat. In the words of S. David Freeman, former White House energy advisor, "Government energy policy has been formulated in Dallas and Houston and rubber-stamped here in Washington."

The oil industry has, of course, been subject to complex government regulations, many of them comprehensible to only a highly trained accountant. Depletion allowances, foreign tax credits, and similar procedures have been standard means to maximize profits. Restrictions have been placed on domestic production of crude oil at times when there was a surplus of oil. For example, starting in the 1930s, "demand prorationing" was applied by a group of Mid-South states to domestic production to "equalize" production, and, as late as the 1960s, Texas wells were restricted to less than 30 percent of their basic allowances. In 1959, President Eisenhower established import quotas on "cheap" Middle East oil to protect United States producers. These were finally abolished in 1973, at about the same time our producers went to 100 percent of capacity.

Such regulations were nothing in comparison to those issued since the 1973–74 Arab embargo. In 1973, the Cost of Living Council dreamed up the "old-oil/new-oil" nightmare that was to enrich so many middlemen. All oil from wells producing before 1973 was considered *old oil* at a controlled price of $5.25 per barrel. Oil from new wells was exempt from controls and could be sold at the market price. Low-productivity *stripper* wells that produced less than ten barrels per day were also exempt. As an example of the way this classification worked out, in 1974 *new oil* averaged about $10 per barrel, almost twice the price of *old oil*—and OPEC oil averaged between $10 and $15 per barrel.

With this price structure, the cost of crude oil to a refiner depended on his relative quantities of old oil, new oil, and OPEC oil. Refiners who were strongly dependent on OPEC oil faced shortages and high prices. They convinced Congress that this was an unfair situation and, in 1974, the Federal Energy Office ordered refiners with crude oil supplies that exceeded the industry average to share their "surplus" crude with their less fortunate competitors. Furthermore, the price of this shared crude was to be the weighted average of the total crude oil costs from all sources of crude. Refiners such as Ashland suddenly found they could reduce their costs by curtailing imports of expensive OPEC oil and buying less expensive oil from crude-rich refiners such as Exxon and Gulf. Supplemental regulations were later issued to soften the impact on the crude-rich refiners, but the entire question of sharing crude caused much bitterness which persists even today. On January 28, 1981, President Reagan ended this controversial "entitlement" program with his executive order decontrolling the price of crude oil.

In 1975, President Ford made an attempt to discourage the importing of OPEC oil by placing a tariff on oil imports. Since Congress refused to do anything except *talk* about the need to reduce imports, he used the authority allegedly delegated to his office by the Trade Expansion Act and, on February 1, 1975, placed a tariff of $1 per barrel on imported oil. This was increased on June 1 to $2; however, in August of the same year, the United States Court of Appeals ruled that President Ford did not have the authority to unilaterally impose oil import tariffs, and—even though this ruling was later overturned by the Supreme Court—the whole question of an import tariff became a dead issue.

The plain truth of the matter is that government attempts to regulate the oil industry have generated confusion, uncertainty, and mountains of paper but have not been for the ultimate benefit of either the producer or the consumer. Particularly bad was the "cheap energy" promotion that produced our rapid growth based on an economy of waste. It can be argued that consumers in the 1950s and 1960s did benefit from low prices for energy, but this will be more than made up by higher prices for future consumers. Oil import quotas imposed when foreign oil was inexpensive only caused a more rapid depletion of our own reserves. The bureaucratic structure that defined such terms as *old oil, new oil, new, new oil,* and *stripper oil* has done little to control the prices of gasoline and fuel oil.

In January 1981, President Reagan finally removed all price controls on crude oil and refined products; however, they are to be replaced by a "windfall profits tax" applied to any increase in the price of oil above the base price established by Washington. Three types of oil will be defined: (1) *Upper-tier oil* from the older fields will be subject to a 70 percent tax on windfall profits. (2) *Stripper oil* from low-capacity wells will be subject to a 60 percent tax on windfall profits. (3) *Newly discovered oil* will be subject to a 30 percent tax on windfall profits.

While this certainly represents an improvement over the previous price-controlled situation, it will still require considerable bureaucratic administration and appears to be an open invitation to fraud. Perhaps it is the best compromise that could be reached; the opportunity to get their hands on many billions of dollars of windfall profits was just too much for Congress to pass up, and we can only hope that this money will be used wisely in the development of alternate sources of energy.

Has the gradual decontrol of crude oil prices proven to be effective? Drilling activity in the United States is now the highest in history. For the first time in some ten years we are finding oil almost as fast as we are using it up. After sharp declines in the early 1970s, domestic production of crude oil and natural gas has now leveled off. Demand for gasoline and heating oils is down because of higher prices, and imports of crude oil are lower. There is absolutely no question that decontrol is proving to be very effective.

We can hope that future legislation will concentrate on those activities that have been effective to date, such as development of new sources of oil and new ways to encourage conservation. Specifically, the following government actions could be very effective in promoting our return to energy self-sufficiency:

1. Allow the price of all crude oil to be set at the prevailing world market price. This would in itself encourage the discovery of new sources of oil and make possible the use of the costly methods of enhanced recovery that can greatly increase production from existing reservoirs. (This was done by President Reagan in January, 1981.)

2. Increase substantially the tax on gasoline—so as to reduce consumption and generate funds to investigate new sources of liquid fuels.

3. Require substantial mileage improvement on all new automobiles. It would also be highly desirable to require development of improved engines that would eliminate the catalytic converters now used for emissions control.

4. Enforce proven means of conservation, such as a national speed limit, closing of service stations on Sunday, energy conservation in homes and in business, elimination of unnecessary lighting, and improved transportation systems.

5. Subject to reasonable environmental controls, open up the many known sources of oil located on government-owned land, offshore, and in Alaska.

6. Recognize that some environmental restrictions will be necessary, but that these must be reasonable and consistent with our need to conserve energy. For example, it is very unlikely that motor vehicle traffic will increase in future years; a gradual decrease is quite possible, in fact, and this should certainly be taken into consideration in adjusting maximum emission standards for automobiles.

What About "Big Oil"?

Surveys have revealed that some 80 percent of the people in the United States attribute at least some of our energy problems to the large monopolistic oil companies that have reaped "exorbitant" profits as a result of the worldwide oil shortage in recent years; many of them believe, in fact, that they even took steps to encourage it. A large portion of the population believes that there is actually no shortage at all, and that the situation has been contrived purely as an excuse to raise prices. Increasingly, we hear remarks to the effect that "they ought to break up the oil companies" or that "the government should take over the energy business." How much truth is there in all of this?

There is no question about the size of the oil companies; in fact of the twenty largest industrial corporations in the United States, ten are oil companies—giants such as Exxon, Mobil, Texaco, Standard Oil of California, and Gulf. Exxon is the largest corporation in the world, with over $40 billion in assets and sales of over $60 billion. Its profits alone are calculated in the billions and have shown *very* healthy increases in recent years.

The large oil companies are *integrated;* in other words, they own and control everything from oil exploration equipment, drilling rigs, oil wells, tankers, pipelines, refineries, and chemical plants to delivery trucks and gasoline service stations. They are *international,* with profitable subsidiaries located throughout the entire free world. They are *diversified,* owning much of the nation's reserves of oil, coal, uranium, natural gas, and other sources of energy. They are also *research-oriented,* doing much of our fundamental research on solar energy, fuel cells, coal utilization, and other potential future sources of energy.

When the average person looks at this tremendous concentration of power, it is not surprising that he or she wonders whether or not it is serving the best interests of the country. Even Congress has at times debated whether *divestiture* (breaking up) of the oil companies might be a good way to promote healthy competition. This could take two forms, either *horizontal* divestiture, which would require them to sell off nonoil activities such as coal and uranium, or *vertical* divestiture, which would involve their splitting into separate exploration and drilling companies, pipeline companies, refining companies, and marketing companies. Numerous government and private studies have been made of the effects of divestiture, however, there is general agreement at this time that it would have little effect on competition within the oil industry or on the ultimate cost of energy to the consumer. More important, it is quite unlikely that divesture would significantly increase production of liquid fuels—and it might well have a harmful effect.

As to whether or not there has been any kind of monopolistic conspiracy on the part of the oil companies to produce high prices and periodic shortages of petroleum products, the record has been somewhat checkered over the years. There appears to be no question that, prior to the 1950s, Middle East oil was produced and sold under a "gentleman's agreement" that limited competition and fixed minimum prices. At that time, eight international oil companies produced practically all the crude oil that went into international trade: British Petroleum, Royal Dutch Shell, *Compagnie Française des Pétroles,* and our Exxon, Gulf, Mobil, Texaco, and Standard Oil of California. These eight companies were joint participants in various oil-producing and marketing consortia with restrictions that severely limited any kind of competition. Many of the restrictions had been entered into with the open encouragement of the American, French, and British governments, and it can be argued that some restrictions were necessary. For example, when Exxon and Mobil acquired an interest in the Iraq Petroleum Company in the early 1930s, they had to subscribe to the 1928 "Red Line Agreement," which obligated consortium members not to compete against each other within the area of the former Ottoman Empire.

When Texaco and Standard Oil of California established Aramco, the availability of abundant low-cost crude oil made it possible for the two partners to offer severe worldwide price competition in refined products. When Exxon and Mobil acquired part ownership of Aramco, they worked out an agreement that changed Aramco into a profit-making business which sold oil to its owners at the world price. This, of course, did act to raise prices and reduce competition in the world oil market.

Beginning in the mid-1950s, other oil companies (both American and foreign) began to acquire oil rights in the Middle East. This greatly reduced the possibility of any meaningful monopolistic restriction of trade, particularly at a time when there was a glut of "cheap foreign oil." By the 1970s, the members of OPEC were gradually obtaining exclusive ownership of their oil, and today it is they who dictate the prices. Under these conditions, any restrictive agreements between the oil companies would be meaningless.

What about competition in the production of domestic crude oil? Here there is absolutely no question that the industry is highly competitive. Thousands of small companies are active in the exploration for oil, and "wildcat" drillers still account for a substantial share of newly discovered reserves. The largest domestic producer, Exxon, controls only one-tenth of our total domestic production. Even the more difficult and expensive exploration of Alaskan and offshore reserves has involved the efforts of dozens of oil companies, and it is virtually impossible that any one of them could obtain a dominant position in these areas.

There appears to be no doubt that transportation, refining, and marketing of oil are competitive. Today's worldwide surplus of supertankers guarantees competition in the shipping of oil. Interstate pipeline companies are already regulated by the government. No one oil company has more than 10 percent of the total United States refining capacity, and the fraction of the retail gasoline market held by any one company is even smaller.

Another area where the oil companies are taking a lot of criticism is in their profits, which have increased astronomically in a period of rising prices and decreased sales. This is not hard to understand if we consider the logistics. Before a single gallon of gasoline gets into an automobile tank, the crude must have been loaded into a tanker in the Persian Gulf, shipped south around Africa and then north across the equator to the United States, pumped into storage tanks, put through various refining operations, blended into gasoline, shipped to a distributor, and eventually delivered to the local service station. All of this obviously requires many weeks, but—at the time when OPEC raises the price of the crude—the price of the refined product is immediately increased, and every drop of oil in storage or in shipment increases in value regardless of the price at which it was purchased. As far as the profit picture is concerned, it is easy to see why the oil companies are doing better in a time of rising oil prices, even though the prices are set by OPEC.

In fairness, it should be pointed out that the oil companies are actually no more profitable on a percentage basis than a good many of the companies listed on the New York Stock Exchange. Viewed from this standpoint, the idea of applying an excess profits tax specifically to the oil industry may seem hard to justify.

It is hard to understand how anyone could seriously suggest that government take over the energy industry. There is no evidence at this time to indicate that such a move would assure greater supplies or lower prices for the consumer. Judging from our experience with specific government operations such as the postal service and with Washington bureaucracy in general, it appears that a government takeover of the oil industry would almost certainly only add to our problems.

OPEC

The admitted purpose of OPEC is to limit production of crude oil and to maintain high prices; in other words, it represents an outright monopoly of the worst type. In fact, it has now reached the point where the price of Middle East crude oil (now over $30 a barrel) bears no relationship whatever to the cost of producing the oil (25 cents a barrel or less). When we look impartially at the way worldwide oil development was achieved, it is hard to blame OPEC for their present policies. Exploitation of Middle East oil was surely an example of foreign imperialism at its worst. Oil is the *only* resource of the Middle East; when it is gone, the entire region will go back to a desert which can support only a few goats and camels. With this in mind, it is easier to understand the desire of the Middle East nations to run their own affairs without interference. It is also easier to understand their continuing mistrust of the West, such as their reluctance to have military bases located in their territories.

OPEC has done us a favor by forcing us to face a situation that should never have been allowed to develop. It is an intolerable position for us to be dependent on imports from unstable regions of the world, even seen apart from the severe financial burden of paying for the imported oil. In 1980, the United States paid some $80 billion for imported oil in spite of substantially lower consumption. In a few years, annual payments to OPEC could become greater than the value of all the shares of stock listed on the New York Stock Exchange, greater than the value of all the farmland in the Midwest. *We have been literally mortgaging the economy of the United States to support our wasteful way of life.*

By now it is obvious to many of us that we should have ended our dependence on Arab oil at the time of the 1973–74 embargo. Oil imports to the United States have dropped somewhat from the peak year of 1977, but this is the result of higher prices for energy rather than any real concerted attempt to reduce consumption. Our other option is to develop alternate sources of energy, some of which are described under "Energy Alternatives." Independence cannot be achieved easily, but every day we wait makes the adjustment more painful.

NATURAL GAS

Like petroleum, natural gas (chiefly methane) is usually found in porous sandstone or limestone sedimentary rocks. In some cases, it is found alone; in others, it is dissolved in liquid petroleum. Since crude oil and natural gas are found in the same geological formations, supplies of natural gas tend to be controlled by the same large companies that produce our liquid fuels. In the United States, the chief reserves of natural gas are in Texas, Louisiana, Alaska, Oklahoma, New Mexico, and Kansas. There are also very large foreign reserves of natural gas, particularly those that accompany the oil fields of the Middle East.

Natural gas was considered at first to be a nuisance and was simply burned off in the oil fields. When the development of high-pressure, long-distance gas pipelines made possible the transportation of natural gas to the cities, it rapidly displaced "manufactured gas" because of its lower cost and higher heating value. By the 1880s, gas lines connected Chicago with the natural gas fields of Indiana. By the beginning of the twentieth century, large quantities of natural gas were becoming available from newly discovered fields in California, Oklahoma, Texas, and Louisiana. With improving technology, it was possible by the 1920s to transport natural gas for distances up to 1000 miles.

Phenomenal growth in natural gas use occurred between 1945 and 1970, when it was the fastest growing major energy source in the United States (see Figure 17). "Gas heats best" became the slogan of the utilities, and millions of Americans agreed. Consumers recognized that natural gas was cleaner and more efficient than coal and also appreciated forgetting about the coal bin, disposal of ashes, and other chores associated with use of coal. Small gas furnaces were installed, and basements became recreation rooms. In 1945, natural gas production was slightly under 4 trillion cubic feet, about one-eighth of our total energy needs; by 1970, production had increased to just under 22

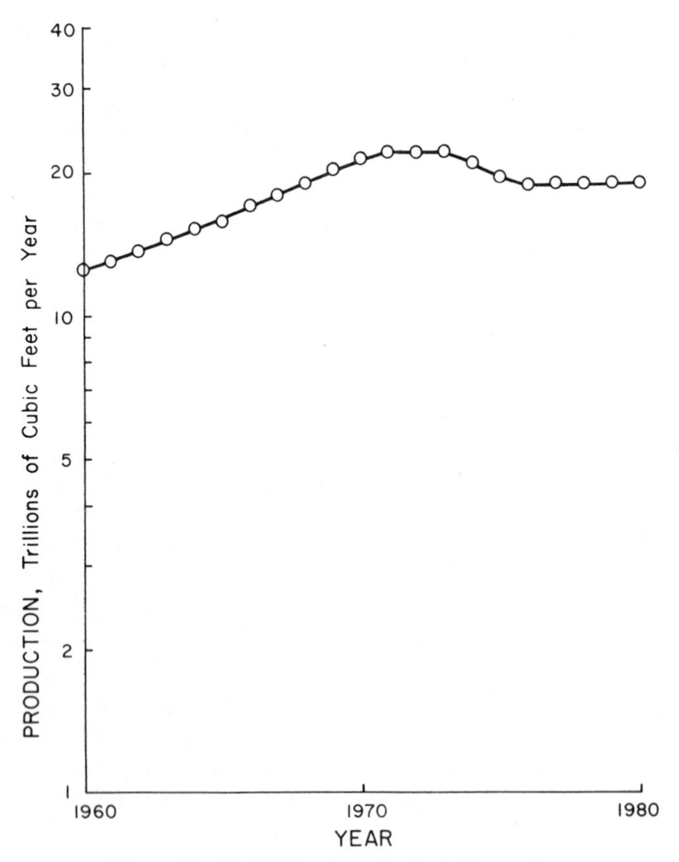

Figure 17. United States production of natural gas.

trillion cubic feet, or about one-third of our energy needs. (During this same time, of course, our total energy demand had doubled.) By the early 1970s, one-half of the nation's homes were heated by natural gas, with most of the remainder heated by oil.

This rapid increase in the use of natural gas produced the same situation that we got into with our crude oil supply. In 1967, proved reserves of natural gas peaked, and they have now fallen by about one-quarter to a value equivalent to only about ten years at our current rate of consumption. In 1973, yearly production of natural gas peaked at 22.6 trillion cubic feet, and it has since decreased to about 20 trillion cubic feet. This has caused increasing shortages of natural gas, particularly noticeable during the winters of 1976–77, 1977–78, and 1980–81, when short supplies necessitated cutoffs to many industries—with resulting loss of business, layoffs, and hardship for thousands of Americans.

In the past, many industrial users of natural gas had "interruptable contracts" which permitted a cutoff during periods of tight supply. Others were simply low on the priority list and found themselves without fuel during stretches of very cold weather when transmission lines could not handle the load. Many users felt they could not take a chance on the future availability of natural gas, in view of pessimistic statements by the federal government, so they switched over to oil or, in some cases, to coal. This obviously placed additional demands on our supplies of crude oil in the post-embargo days when things were already bad enough. Not surprisingly, when natural gas became more available in the late 1970s, most of these users refused to switch back.

From time to time now it appears that we have a surplus of natural gas (commonly known as the "gas bubble"), and the Department of Energy has been encouraging, even urging, industrial users and utilities to convert from oil to gas. This temporary surplus—and that is what it is—is the direct result of the conversion by low-priority users from natural gas to oil or coal and also of higher prices for natural gas, which have acted to increase supplies and encourage conservation by consumers. There is much confusion about the entire natural gas picture, which has resulted in considerable wrangling on the part of industry, Congress, and the government energy experts. At one point during the 1978 congressional hearings, former Energy Secretary James Schlesinger commented, "I understand now what hell is. Hell is endless and eternal sessions of the natural gas conference." Even after the final passage of the National Gas Policy Act, one industrialist remarked that Secretary Schlesinger had seen only the beginning of the debate!

Regulation of Natural Gas

Many of today's problems with natural gas are a direct result of over fifty years of erratic government regulation. As public utilities, the early natural gas distribution companies were subject to state and municipal regulation as "natural monopolies" in each area. When natural gas pipelines began to cross state boundaries, the price charged to consumers in one state was determined by the price 1000 or more miles away in another state. Regulation by the individual states became impossible, so that in 1938 the Federal Power Commission (FPC) was created to assure that prices charged by the interstate pipeline companies to local gas distributors were "just and reasonable." (Later, when the Department of Energy was created, the FPC became the Federal Energy Regulatory Commission, or FERC as it is now known.)

Authority of the FPC was strengthened in 1954 by a historic Supreme Court decision in the case of *Phillips Petroleum Co. vs. State of Wisconsin.* The Court ruled that gas consumers needed protection from price increases that were being passed through by the pipeline companies, and that hence the FPC had the obligation to regulate the price of natural gas in the oil fields, the so-called *wellhead price.* Unfortunately, this decision failed to state how a "just and reasonable" price could be obtained. It also created two different markets for natural gas, an *interstate* market controlled by the FPC and an *intrastate* market—since gas that did not cross state boundaries would be free from price controls.

At first, there did not seem to be a real problem in applying the FPC regulations. Natural gas in 1954 was so inexpensive that some producers still just flared it, and few wells were drilled primarily to find gas. The average price for the natural gas that found its way to the market was only about one-fifth the price of fuel oil having an equivalent heating value. Interstate prices continued to rise moderately just as they had in the past. With ample supplies, there was essentially no difference in price between interstate and intrastate markets.

The situation began to change in the 1960s when declining reserves signaled the approaching end of cheap and abundant natural gas. In 1967, the FPC started to allow higher prices for newly discovered natural gas. At the same time, the price of natural gas sold within the state started to rise, so that by the early 1970s the price of intrastate gas was over twice that of regulated interstate gas. Producers naturally preferred to sell for the best price they could get, and by the 1970s about 90 percent of the newly discovered gas was going into the intrastate market—much of it used as a fuel by the electric utilities in place of coal. Lee White, former FPC chairman, termed this wasteful process "a national scandal." Needless to say, the fortunate intrastate customers had plenty of natural gas even during the critical winter of 1976–77!

By the late 1970s, the FPC finally recognized that higher prices were necessary to guarantee the supply of interstate gas, and the price was allowed to rise until it was not too far below that of the uncontrolled intrastate gas. In 1978, Congress passed the Natural Gas Policy Act, which provides a guide for the eventual deregulation of natural gas prices. The wellhead price of all newly discovered natural gas will be allowed to rise each year for ten years, at which time it will be deregulated. To eliminate the two-tier market, the wellhead price of newly discovered natural gas consumed in the producing state will also be controlled. The price of "old" interstate gas will be unchanged, but higher prices will be allowed as contracts for old gas expire. To further complicate the picture, price rises are not automatic; producers must renegotiate each contract to obtain any price increases permitted under the Act.

One thing is certain: there will be continued controversy over natural gas. The "old-gas/new-gas" pricing will be difficult to enforce and is an open invitation to fraud. It should also be noted that the government has assigned the following priorities among the users of natural gas: (1) residential users, (2) industries using natural gas for chemical feedstocks, (3) industries with gas boilers, and (4) electrical utilities. With this priority system, it is not surprising that the large industrial users of fuel continue to be skeptical about future supplies of natural gas, and it is unlikely that they will be willing to switch from oil or coal.

Another factor that complicates the whole natural gas picture is the difficulty of assigning an actual cost to natural gas. About one-quarter comes from oil wells, where a purely arbitrary allocation must be made between the cost of the gas and the cost of the oil. Many natural gas wells are discovered during oil exploration, with the natural gas an unexpected by-product. One thing on which there seems to be general agreement is that while modest increases in price will stimulate production of natural gas, once above a certain price, further increases have little effect on production.

It is also generally agreed today that there is enough domestic natural gas to keep us going at the present rate of consumption for another twenty or twenty-five years, particularly if we get a little help from Canada and Mexico. After that, we will have to rely increasingly on substitutes.

Liquefied Natural Gas (LNG)

One of the major problems in the distribution of natural gas has been the winter heating load, which places much higher demands on the pipelines and on the distributors. Two methods were used to even out the demand. Some of the distributors negotiated low-price interruptable contracts with electrical utilities, which permitted them to use natural gas as fuel during the summer months when demand was low. Others built large *gasholders* which were quite expensive but could be used to augment supplies during the winter months of high demand.

It has been known for some time that storage of natural gas can be considerably increased if the gas is first liquefied, since some 600 cubic feet of natural gas yield less than 1 cubic foot of liquid. To do this requires cooling the natural gas to very low temperatures since methane (its chief constituent) boils at 164 degrees centigrade below 0 at ordinary atmospheric pressure. In addition, the LNG must be stored in well-insulated tanks that are continuously cooled to prevent a build-up of pressure which might lead to an explosion.

The storage of LNG was investigated as early as the 1930s and was used by gas distributors during the early 1940s. On October 20, 1944, in Cleveland, Ohio, a cylindrical storage tank containing 147,000 cubic feet of LNG ruptured and released the LNG into an enclosed area. Some of this liquid overflowed into storm sewers of the adjacent neighborhood, and subsequent fires and explosions damaged an area extending for more than a quarter of a mile from the tank. Flames were reliably reported to have reached 2800 feet in height; 135 lives were lost and hundreds were injured, with property damage in the millions. This tragic accident effectively stopped the use of LNG in the United States for the next twenty years.

Yet the sight of bright orange flares burning natural gas in overseas oil fields was too much to ignore, and by the 1960s LNG had become an important material of commerce. Some 150 million cubic feet of natural gas were being liquefied each year in Algeria alone, and specially constructed ocean-going tankers were shipping it to England and to northern France. The first LNG imported to the United States reached Boston in 1974, and many natural gas distributors regarded it as the salvation of their industry.

To date, our government has been very reluctant to encourage any growth in LNG imports beyond the present very modest level. First, there is the question of safety. An accident at LNG storage facilities or to an LNG tanker entering port or unloading fuel could produce catastrophic fire and explosions. It is true that there have been no serious accidents since the Cleveland explosion; however, very serious accidents have occurred with other liquefied gases such as propane and propylene, both less hazardous than LNG. Many experts claim that the special design features of LNG storage facilities make it no more hazardous than gasoline—many others claim that handling of LNG imposes great risk to the public.

In addition, there is the question of the cost of LNG. Because of its very high refrigeration, handling, storage and delivery costs, the LNG business requires an initial investment of some $2 billion. Under such conditions, LNG cannot compete economically with domestic natural gas. Recognizing this, the Algerian government accepted very low prices for LNG in their first contracts with American firms. Almost all gas experts feel that the price will increase sharply in the future and will eventually be close to the equivalent OPEC price of crude oil. (It is known, in fact, that OPEC is presently looking into possible ways to control the world market for LNG.)

At present, the United States imports about 5 percent of our natural gas from Canada by pipeline. Negotiations for much greater quantities of natural gas from Mexico broke down in 1978 over price, but it is likely this will eventually be resolved. In the meantime, it is clear that the importation of large amounts of LNG would place us in the same position of foreign dependence that we now have in the case of crude oil. The only difference would be that radical Algeria rather than conservative Saudi Arabia would control supplies. It is easy to understand our reluctance to get into that trap!

While the total amount of energy available from the world's proven reserves of natural gas is about the same as that available from the proven reserves of crude oil, prospects for discovering additional reserves of natural gas are better than prospects for discovering additional crude oil. In addition, we know that less than one-third of the OPEC natural gas is being marketed at present; some is reinjected into the wells, and over one-half of it is still flared. As the energy shortage continues to escalate, the world will undoubtedly begin using a greater fraction of its natural gas reserves.

ELECTRIC POWER

Electricity is a form of energy that can be easily converted to work or heat. It consists of the flow of minute negatively-charged particles (*electrons*) through a conducting material. Electricity is produced in a power plant and distributed to consumers by means of electric transmission lines. The amount of resistance offered to the flow of electrons—that is, to the electric *current*—is important in the selection of a conductor. Because of their low resistance, copper and aluminum are generally used for transmission lines so as to minimize "line losses."

The rate of the flow of electrons is measured by the *ampere,* named after André Marie Ampère (1775–1836), a French physicist who developed many of the basic concepts of electricity and magnetism. The potential of the current is measured in *volts,* named for the Italian physicist Alessandro Volta (1745–1827), who created the first electric battery. Electric potential can be compared to gravitational potential. We have seen that water in a reservoir at the top of a hill can be run through a turbine at the bottom of the hill to produce work; similarly, an electric current available at high voltage can be run through an electric motor to the *ground* (zero voltage), thereby producing work.

The work or energy that results from the flow of an electric current is measured in *watts,* named after the Scottish inventor of the modern condensing steam engine, James Watt (1736–1819). The product of the amperes and the volts gives the watts: an electric broiler that uses 10 amperes of current at ordinary 110-volt house current consumes 1100 watts of electricity. Both the watt and the kilowatt (1000 watts) measure the rate at which energy is consumed. The total amount of energy consumed is usually given in *kilowatt-*

hours (kwh), which measure both the power and time of use. If the broiler is used for 1 hour, our electric bill will include 1.1 kilowatt-hour of electricity.

Commercial development of electricity was based largely on the early work of Michael Faraday (1791–1867), an English chemist and physicist who was director of the Royal Institution of Great Britain for many years. Faraday found that a wire carrying an electric current rotated continuously about a permanent magnet. This discovery stimulated practical inventors to build machines called *dynamos* (now known as *generators*) to convert mechanical energy into electrical energy. In 1873, it was found that a dynamo could be reversed and used as a motor to convert electrical energy into mechanical energy. In 1879, the American inventor Thomas Alva Edison (1847–1931) developed the carbon-filament incandescent electric light bulb. In 1880, there was a well-publicized electrical exhibition at the Crystal Palace near London. The first electric "tram" car was demonstrated in Paris in 1881 and was widely adopted in the United States by the close of the century.

The first Edison central station distribution system was the Pearl Street steam-engine system which began operating in New York City in September, 1882. For the next ten years or so, the industry was quite small; early direct-current generating plants averaged only 200 or 300 kilowatts and the distribution area was limited to about 1 square mile. Each small town that wanted electricity had its own plant, and large cities required many power plants to supply their needs.

By the end of the nineteenth century, the new electric utility industry had begun to shift to alternating current, which could be distributed much more easily. Commercial and industrial establishments soon saw the advantages of electricity, and its use spread rapidly in the early twentieth century. In 1902, there were approximately 3600 electrical generating systems; by 1917, this had grown to 6500. Following World War I, there began a period of mergers and consolidations which has continued to the present. Today, slightly over 200 privately owned electric utility companies generate over 75 percent of the total United States production. The remaining electric production is roughly divided equally between the federal government systems (such as the Tennessee Valley Authority and Bonneville) and some 2000 publicly owned systems operated by states, counties, municipalities, and electric cooperatives.

Electric power is something all of us take for granted in the United States, something we depend on at any time of the year and at any hour of the day or night. It heats our homes, either directly through electric heat or indirectly by driving our furnaces. It pumps the water we use for drinking, washing, and sanitation. Often it cooks our meals. It provides our light, amuses us through television, and powers the host of electrical devices that make life easier and more enjoyable.

Electricity is equally important in the work place, where it provides each worker with extra "hands" to lighten the load. Many of our industrial chemicals (such as aluminum, chlorine, phosphorus, and acetylene) are produced by electrolytic processes. Modern offices are dependent on electric typewriters, recording equipment, and computers. Without electricity, our entire economic system would collapse.

Loss of electricity for even a few hours can cause severe problems. On two occasions, blackouts in New York City have resulted in wide-scale rioting and looting. In the winter of 1973, a severe ice storm led to widespread power failure in central Connecticut; without heat for up to a week, many homes were evacuated, water pipes froze, and much damage resulted. Try to imagine life without electricity for a few days or a few weeks—bare survival would be possible in a suburban or rural area, but even animal existence would be doubtful in the brick and concrete jungles of our large cities.

The electric utilities are encountering many problems today, several of their own doing but many are a direct result of the laissez-faire attitude that has governed the rapid growth of the industry during most of its first 100 years. There is considerable uncertainty regarding the future demand for electric power and the extent to which new generating facilities will be needed. It is difficult to locate new plants because of environmental regulations and vociferous local opposition—particularly to nuclear plants. Many of us question the fairness and usefulness of rate structures which reward large consumers with lower rates, and vocal consumer groups oppose any rate increases. Caught between decreased profitability and higher interest rates, the utilities are finding it more and more difficult to obtain funding for the enormous cost of new capital equipment.

These problems must also be viewed against the background of increasing fuel costs. The electric utilities use more than 30 percent of the total amount of fuel consumed in the United States, and, of course, any increase in the cost of fuel is immediately reflected in the cost of producing electricity. Also, the utilities are under pressure from the federal government to reduce their use of oil by converting, or reconverting, to coal—a very time-consuming and costly project in itself. With still higher costs and possibly reduced consumption of electricity in view, it is not hard to understand the industry's deep concern for the future.

DAILY AND YEARLY POWER-PEAKING

To think constructively about the problems involved in the generation and consumption of electricity, we must distinguish between the *energy consumption* or *total use* of electricity as measured in kilowatt-hours and the *demand* or *rate* at which the electricity is used. For example, the consumption of 24 kilowatt-hours of electricity can be achieved by using 1 kilowatt for a period of 24 hours, by using 24 kilowatts for 1 hour, or by any other combination in which the product of rate and time yields the number 24. For the average residential consumer, the cost of a given number of kilowatt-hours is the same regardless of the rate at which the electricity is consumed or the time of day or night.

For the large commercial or industrial user, there is both an *energy charge* and a *demand charge.* The energy charge is similar to that paid by a residential consumer, except that the price per kilowatt-hour is usually lower for these larger users of electricity. The demand charge measures the maximum rate at which the electricity is taken and in many cases substantially raises the total electric bill. Large users can reduce their costs by spreading out their usage of electricity rather than by taking large amounts over a short period of time.

Since an electric utility company must have sufficient generating capacity to meet the maximum demand placed on its system, its chief problem is the very wide variation in demand. Figure 18 provides an example of the demand experienced by Northeast Utilities (the largest utility company in Connecticut, with a fairly balanced residential-commercial-industrial load) during a typical winter week. Note that the maximum demand, or *peak,* of 4 million kilowatts

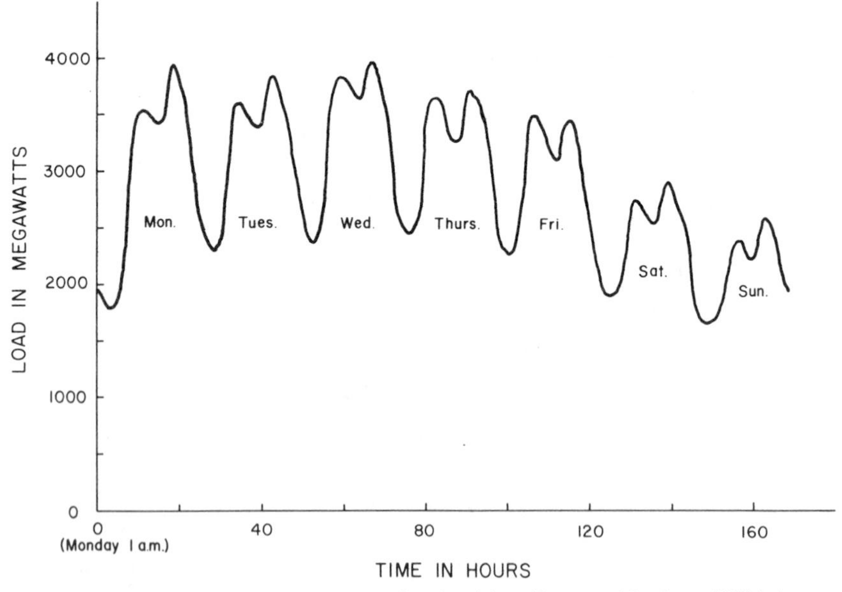

Figure 18. Typical weekly demand for electricity. (Courtesy Northeast Utilities)

occurs during the afternoon rush hour on Monday through Thursday, with a second slightly lower daily peak occurring during the morning rush hours. On Friday, there is a slackening of demand, and during Saturday, Sunday, and Monday night the demand falls below 2 million kilowatts—less than half the maximum demand.

The situation is further complicated by a yearly power-peaking which is superimposed on the daily pattern. Consumption of electricity is lowest during the spring and fall months; in summer, there is increased demand resulting from the air-conditioning load and, in winter, there is a second peak resulting from the electric-heating load. An electric utility must plan to have enough generating capacity to supply the maximum demand that can result from the piling up of these daily and yearly peaks, even though much of this total capacity will only be used during a few hours of the entire year.

In a state such as Connecticut, where electric heat has been widely promoted by the utilities themselves, the maximum generating capacity is determined by the winter peaking which occurs at about 5 p.m. on a cold day in January or February. In warmer regions, such as the Southwest, the maximum generating capacity is determined by the summer air-conditioning load. Where there is a very high industrial load, such as in the Chicago or Pittsburgh areas, daily

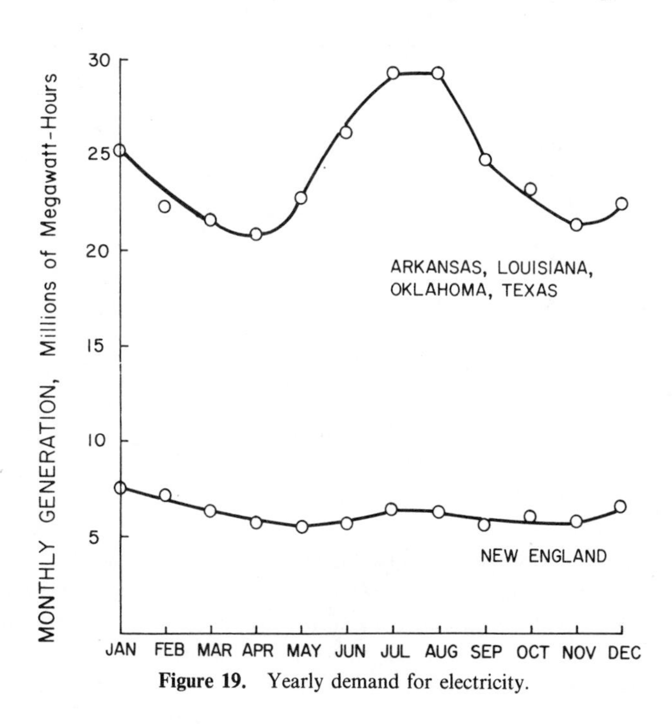

Figure 19. Yearly demand for electricity.

peaking is much more pronounced and demand during the day may be three times or more that at night. In more temperate regions, such as San Francisco and the Pacific Northwest, daily and yearly power-peaking are much lower. Figure 19 provides an example of the yearly demand for electricity.

This unbalanced demand is a significant factor in determining the cost of electricity, since it is obviously inefficient to use expensive generating equipment on only a part-time basis. Northeast Utilities is presently constructing a fourth nuclear plant in Connecticut in anticipation of the increased demand expected in the 1980s. The cost of this plant, currently estimated at about $2 billion, will sooner or later be passed on to the consumer. It is entirely possible that the need for this very expensive new facility could have been postponed or even eliminated by such steps as the shifting of some industrial demand to off-peak hours, minimizing the use of air conditioning, and outlawing any new installations of wasteful electric heating.

CYCLING OF GENERATING FACILITIES

How does an electric utility company handle the cyclic demand for its product? First, any large utility will have many generating facilities of different types, each producing electricity at different cost. The most cost-efficient of these are designated *base-load* plants and operate continuously around the clock. These normally include the hydro plants, the nuclear plants, and the most cost-efficient of the fossil-fuel plants as needed to supply the minimum system demand. Figure 20 illustrates how electricity is produced in the United States.

As the demand starts to rise, for example on a Monday morning, other units will be gradually brought into service to handle the added load—beginning with the next most cost-efficient plants. To take care of the morning and evening rush hours, it may be necessary to resort to the most costly or *peaking* units; these plants require petroleum or natural gas and can be started up or shut down easily as needed. As the demand starts to let up at the end of the afternoon, the process will be reversed until eventually the system is back to base load. The whole procedure is designed to minimize the cost of electricity to the consumer.

In areas such as the industrial Midwest, where most of the electric generation is either nuclear or coal-fired, the cyclic demand can present a challenge. Large coal-fired plants respond very slowly to changes in demand and must be operated at perhaps one-half or one-third of normal power at night to be ready for peaking loads the next day. Utilities are also very reluctant to shut down nuclear reactors or even to reduce their power, because these plants

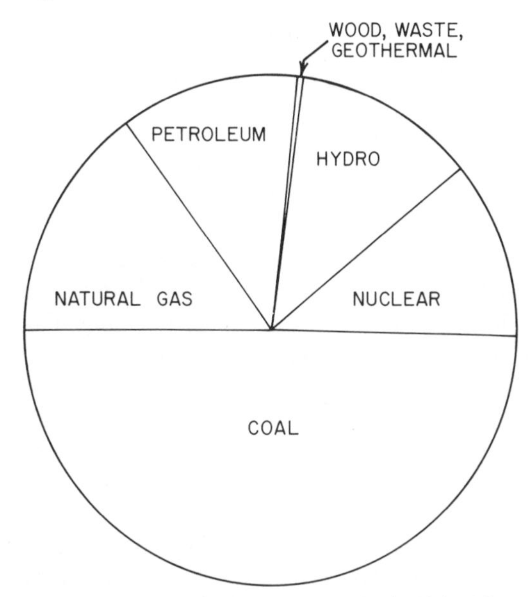

Figure 20. How electricity is produced in the United States.

require many hours to return to full power. To avoid the necessity of further reducing the system load, there have actually been times when Midwest utilities have been willing to give away power during off-peak hours!

In other areas, the cyclic demand is no real problem, particularly where power-peaking is not too great, or where much of the base load can be used for *pumped storage* as in upstate New York or New England. This process involves constructing at the top of a mountain a large reservoir that is connected through turbines to a river at the base of the mountain. When demand is low, water is pumped from the river up to the reservoir and stored there as potential energy until needed. During peak loads, the flow is reversed and the pumps operate as turbines to produce electricity.

Pumped storage gives considerable flexibility to a utility company. For example, the Northfield Mountain facility on the Connecticut River uses base-load nuclear power during nights and weekends to pump water; to meet peaking demand or emergencies, it can be brought to 1 million kilowatts in less than 1 minute. While this involves some loss of power (about 60 percent of the energy used to pump the water is actually recovered again as power), it eliminates the need for new generation facilities and does not consume critical fuels such as petroleum and natural gas.

A minor environmental problem in connection with pumped storage is the large reservoir that must be provided and the continuous fluctuations in water level which prohibit any possible recreational use. Other methods of storing electricity under investigation include pumping gas under pressure into a large underground cavity so that it can be run through conventional turbines under high pressure to generate electricity to meet peak demands. These newer methods are limited, however, in the amount of energy they can store, and it appears very unlikely that there will be any practical substitute for pumped storage in the foreseeable future.

In addition to the challenge of cyclic demand, all utilities must allow for a certain amount of "down time" while individual facilities are under repair, maintenance, or in the case of nuclear plants, refueling. For this reason, a utility company normally maintains at least 20 percent excess generating capacity and, whenever possible, schedules maintenance and refueling for the spring and fall months when yearly power demand is at a minimum. This has become more difficult with the trend toward larger units, and some utilities have run into real trouble when several key units went out of service at a time of high demand.

To minimize such disruptions, interconnections have been established that permit an exchange of power between nearby states. The New England Power Exchange, for example, coordinates all power generation for the entire region. While these interconnections help to avoid power interruptions and permit the most efficient units to be operated at any given time, thereby reducing costs and conserving fuel, they also magnify the potential for disaster—as in November of 1965, when a failure in a hydroelectric station in Canada blacked out almost the entire northeastern United States for several hours.

HYDROPOWER

Electricity is generated by four common methods: water turbine (*hydro*), steam turbine, gas turbine, or diesel engine (*thermal*). Of these, hydro is by far the least expensive and the best from an environmental standpoint. At one time, all hydropower was harnessed by building a dam to divert water to the top of a water wheel; buckets on opposite sides of the wheel alternately filled and emptied, turning the wheel by the action of gravity to produce mechanical energy. Today's hydroelectricity is produced by a conventional water turbine; water under high pressure rotates blades mounted on a shaft directly connected to a generator, which converts the mechanical energy to electrical energy.

A modern hydro plant is situated near the base of a reservoir or waterfall. Water is admitted to the inlet of the turbine either through a tunnel at the bottom of the dam or through a pipe if the turbine is located downstream from the dam. Discharge water at atmospheric pressure leaves the turbine and flows downstream. There is no need to worry about exhaust gases, scrubbers, precipitators, radioactive fuel elements, and similar problems encountered in thermal methods.

The quantity of electricity produced depends on two factors: (1) the rate at which water flows through the turbine and (2) the *head*, or difference in elevation between the turbine and the water level at the top of the reservoir or waterfall. Since hydroelectric plants convert potential energy directly into electricity (in contrast to thermal methods which convert heat into electricity), their efficiency is quite high, ranging up to 90 percent for the best installations. Some hydro plants are very small, with some commercially operated units producing as little as 200 or 300 kilowatts; others are among the largest power stations in the world. There are several plants of about 2 million kilowatts in the United States, three of more than 4 million kilowatts in Soviet Russia, and the Itaipu Dam on the Brazil-Paraguay border will have an eventual capacity of almost 13 million kilowatts. Some nations, such as Brazil, Canada, Chile, Norway, Sweden, and New Zealand, obtain most of their electricity from hydropower. In the United States, slightly over 12 percent of the total electric generation comes from hydro.

Hydroelectric plants do have some adverse social and environmental aspects. They often deny access to the spawning grounds of important migratory fish, such as salmon, although this can be avoided by use of fish ladders around the dam. Large reservoirs in some cases cover the best farmland and may flood small towns or villages. Stream conditions can be substantially altered downstream from a dam, particularly by the rapid fluctuation of the water level often encountered during intermittent power generation. Underground seepage from reservoirs can raise the surrounding water table, sometimes dissolving subsurface salts that may impair the fertility of the soil. In addition, there is always the potential for a disastrous flood if a dam upstream from a populated area should suddenly give way.

The availability of hydroelectric power can vary considerably over the period of a year, normally being at a maximum during the early spring and at a minimum during the warm summer months. If a reservoir is also used for irrigation or for maintenance of a navigational channel, not all of the water will be available for power production and there may be limitations on its use. If a reservoir is used for flood control, it must be kept low during periods when floods may occur; this reduces the available head and may also restrict the amount of water that can be used later for power production.

Although built primarily for irrigation purposes, the Aswan High Dam in Egypt has a hydroelectric production of about 2 million kilowatts and provides an interesting example of the unforeseen problems that may result from the construction of a large reservoir. Since completion of the dam, the change in flow of the Nile River has practically wiped out fisheries in the eastern Mediterranean; along the lower Nile, fertility of the soil is gradually decreasing without the nutrients formerly carried down by periodic flooding; the wide network of irrigation channels has created ideal conditions for the spread of certain small snails which transmit the serious parasitic disease known as *schistosomiasis,* or *bilharzia;* and, finally, current figures show that increased food production on this newly irrigated land is proving insufficient to support the growth in Egypt's population which occurred during construction of the dam.

Hydroelectric power is usually considered to be a renewable source of energy, since it depends only on upstream precipitation. It should be pointed out, however, that all reservoirs will eventually fill with silt carried from upstream and deposited over tens or hundreds of years—raising some question as to the long-range future of much of our hydropower capacity.

The availability of hydroelectricity is, nevertheless, becoming more and more important to us as the cost of fossil fuels continues to increase. While in certain parts of the world such as Asia, Africa, and Latin America, the developed hydropower represents only some 5 percent of the potential, most of this undeveloped hydro is located far from electric-using markets. In the United States, roughly half the potential hydropower has already been developed, including practically all the easily developed sites. In addition, further development of hydropower in this country has come under increasingly vociferous environmental attack. It is obvious that we shall soon face very unpleasant choices; for example, should we continue to preserve the magnificent view of Niagara Falls as we pay ever higher prices for imported oil, or should we continue to develop our hydro potential even if it turns Niagara into a rocky cliff?

THERMAL GENERATION OF ELECTRICITY

Roughly 85 percent of the electricity generated in the United States comes from the conventional steam turbine. High-pressure steam is produced in a boiler by burning a fuel such as coal, oil, or natural gas, or by using a nuclear reactor as a source of heat. The high-pressure steam is run through a steam turbine to generate electricity, and the low-pressure exhaust steam is sent to a condenser where it is liquified and recycled back to the boiler or reactor. In this process, roughly two-thirds of the heat energy of the fuel is "dumped" as waste heat. Since tremendous quantities of cooling water are needed for the steam condenser, close proximity to a very large body of water is the first requirement in the location of any type of thermal power plant.

Coal supplies over half of the total thermal power in this country. A large coal-fired power plant requires thousands of tons per day and produces hundreds of tons of ash per day. As we have learned, there are several substantial disadvantages associated with the use of coal: (1) It is necessary to use costly filters and electrostatic precipitators to remove dust and fly ash from the exhaust gases. (2) Most of the sulfur in the coal burns to form sulfur dioxide gas, which leaves with the exhaust gases in most plants. Since this causes acid rain downwind from the power plant, extremely expensive scrubbers may soon be required to eliminate this problem. (3) The coal ash (plus effluents from scrubbers if they are required) represents a very substantial solid waste disposal problem. (4) Smog-producing nitrogen oxides (a mixture of nitric oxide and nitrogen dioxide) released with the exhaust gases may develop into a significant health hazard in some areas where there is increased use of coal in large power stations.

Under President Carter's energy program, utilities have been asked to convert or reconvert oil-fired boilers to coal wherever possible. This is meeting considerable opposition from the industry. In addition to the substantial amount of production that will be lost while any such changes are made, they point out that the new equipment will be very costly (and scrubbers would add to this cost), and that many of their present units no longer have facilities to receive coal by rail or barge or to dispose of the large quantities of ash produced.

Oil and natural gas each account for about one-sixth of the total thermal generation of electricity in the United States today. Ten years ago, these fractions were each slightly over one-fifth. We are evidently making some progress in reducing our dependence on these fuels, but it is very slow going!

Much of the current use of oil and natural gas is for intermittent peaking power and cannot be eliminated without the construction of very expensive new coal-fired or nuclear plants. Some use is seasonal; for example, natural gas is commonly used in the summer when there is no residential heating load and ample supplies are available. Under these conditions utilities have sometimes entered into low-priced "interruptable" contracts with gas suppliers. Now, however, they are becoming very reluctant to make long-term commitments based on the availability of natural gas—despite assurances from the federal government that there is plenty of natural gas around. There have been shortages in the recent past, and the utilities are well aware that they are at the bottom of the priority list in any possible future shortage of natural gas.

Rather than using expensive large-scale nuclear or coal-fired generating facilities, we have seen that the utility companies often resort to small peaking units during short periods of high demand for electricity. Two general types of unit are used, the *gas turbine* and the *diesel engine.* Both burn expensive fuel; however, they operate for short periods of time and eliminate the huge investment cost of providing this power by means of large-scale facilities.

Although they can be operated on natural gas, gas turbines normally burn a fuel oil similar to residential heating oil. The oil is simply burned with air under pressure and run through a turbine to produce the electricity. Exhaust gases pass to a stack and are discharged to the atmosphere. Diesel engines are another common means for generating electricity during short periods of high demand. These are merely very large diesel engines similar to those that power ocean-going ships. Burning conventional diesel fuel, they are directly coupled to a generator to produce electricity and their exhaust gases pass to a stack.

President Carter's energy program essentially banned construction of any new oil-fired plants. The use of oil is being kept to a minimum already by the utilities because of its high cost and will continue to drop as older units are phased out.

Efficiency in Thermal Generation

A major consideration in the generation of electricity is the efficiency of the process used—that is, what fraction of the energy of the fuel is actually converted to electrical energy? The transformation of heat into work is an example of the transformation of energy from one form to another and is therefore governed by the laws of thermodynamics. When these laws are applied to processes involving the production and utilization of heat energy, an interesting and important conclusion is reached.

When other forms of energy (such as electrical, potential, or kinetic) are transformed to heat or work, as in hydroelectric plants, we always find that the process can be carried out with an efficiency close to 100 percent. When we go in the opposite direction, however, converting heat to work (thermal generation), we find that the maximum efficiency possible is only about 30 to 40 percent. The fact that heat represents a less valuable form of energy, not as useful as the other forms, is critical in the operation of any electrical-generating facility.

The thermodynamic principles involved in the operation of an engine or steam turbine to produce work were first developed by the French engineer Sadi Carnot (1796–1832). The essential conclusion is that the maximum possible efficiency of any engine or turbine depends only on the *absolute temperature* of the fluid entering or leaving the engine. (Absolute temperature is the number of degrees above *absolute zero,* which is −273 degrees centigrade or −460 degrees Fahrenheit.) Thus, if steam enters a turbine at a typical temperature of 260 degrees centigrade (533 degrees absolute) and is exhausted at 40 degrees centigrade (313 degrees absolute), the maximum efficiency is (533–313)/533, or about 40 percent. The actual efficiency will always be lower than this, but what we must remember is that, in a typical thermal electrical power plant, only about one-third of the heat energy of the fuel is converted to electricity —the remaining two-thirds are simply dumped as waste heat.

Since practically all utility plants dump their waste heat to an adjacent large river, lake, or ocean, this has often raised the question of possible thermal pollution of the environment. Utilities are coming under increasing attack by environmentalists who feel that even small temperature changes in large bodies of water may be harmful. Even though every thermal power plant has this problem, much of the attack is directed at the nuclear facilities because of their large size and inherently lower thermal efficiency. While it is possible to install cooling towers that effectively discharge the heat to the atmosphere rather than to water, these towers are extremely expensive in themselves and require large amounts of power for operation. They do not represent a satisfactory solution to the problem of thermal pollution.

Cogeneration

We have seen that the efficiency for changing heat into work can be increased by lowering the temperature of the fluid leaving a steam turbine or an engine. For this reason, utilities operate their exhaust steam condensers at the lowest temperature they can achieve using the cooling water available to them. Under these conditions, the exhaust steam is at such a low temperature that it has no value even though it accounts for some two-thirds of the heat energy of the fuel that has been burned.

It is perfectly possible, however, to run a steam turbine under such conditions that the steam is exhausted at a pressure high enough to permit the steam to be used for domestic or industrial heating. In this way, much of the heat energy in the exhaust steam can be recovered. Even though the electric production is reduced, the total amount of energy recovered (electric energy plus exhaust heat) is perhaps double the amount of energy recovered when only electricity is produced.

This process, known as *cogeneration,* can save substantial quantities of energy while at the same time reducing thermal pollution of both air and water. At one time it was used by United States industry to supply some 30 percent of its electric needs, but this figure has been continuously dropping. Today, only about 8 percent of our industrial electricity comes from cogeneration. There are several difficulties: (1) Skilled operating personnel are required for cogeneration, so that the cost of labor has tended to offset any fuel savings. (2) Cogeneration requires close cooperation between the industry and the utility for the sale of surplus electricity and the purchase of backup power; in particular, electricity produced during off-peak hours may have little value to the utility. (3) Industry is understandably afraid that cogeneration may open up further regulation by state and federal bureaucrats.

Another possibility, used in Europe and known as *district heat,* involves the direct sale of exhaust steam by a utility for domestic heating. This requires the location of a small electric power station in a medium or large city and the distribution of exhaust steam by pipeline to buildings in the vicinity of the plant. The problem here is that a relatively expensive backup system is needed, since any power plant must occasionally be shut down for maintenance or repair. The cost of the electricity is higher because of the lower thermodynamic efficiency and higher costs associated with a small-sized unit; in addition, there are costs associated with the distribution and metering of the steam. Even though district heat offers a way to conserve energy, the total cost may well be higher than that of the conventional systems.

While increased use of cogeneration should certainly be encouraged by providing financial incentives and by modification of regulations that control the sale of surplus power to the utilities, it is very unlikely that cogeneration will help us very much in the immediate future.

NUCLEAR POWER PLANTS

Nuclear energy has become an increasingly important source of energy for the United States during the past ten or fifteen years and now represents about 13 percent of our total thermal generation of electricity. President Carter's energy program assumed continued reliance on nuclear energy to furnish an even higher percentage of our energy needs. In some regions of the country, such as New England, nuclear power is by far the cheapest source of thermal power. Despite these facts, the entire future of nuclear power in the United States is now very much in doubt.

The story of nuclear power goes back to development of the atom bomb during World War II. In 1944, General Leslie Groves appointed a committee to look into possible peace-time uses of atomic energy, and they came up with a multitude of suggestions, ". . . principally along the lines of the use of nuclear energy for power and the use of radioactive byproducts for scientific, medical and industrial purposes. . . . There is a good probability that nuclear power for special purposes could be developed within ten years. . . ." It was further recognized that some form of government control and support in the field of nuclear energy would have to continue after the war ended.

In 1946, Congress established the Atomic Energy Commission, and President Truman appointed David Lilienthal as its director. According to the preamble to the Act, their mission was defined as follows: "The development and utilization of atomic energy shall be directed toward improving the public welfare, increasing the standard of living, strengthening free competition in private enterprise and promoting world peace. . . ."

In the early 1950s, the proponents of atomic energy really shifted into high gear. "We will eliminate the ills of humanity. In the brave new world of tomorrow, atomic energy will make the deserts bloom. Electricity will be so cheap you won't even have meters on your houses." (Of course, they forget one small item, that fuel cost was only about one-sixth of the total delivered cost of electricity.) In 1951, an advisory committee recommended future cooperation between the electric power industry and the Atomic Energy Commission. In 1954, the Atomic Energy Commission announced that $242 million would be allocated to the development of power reactors, with the goal of achieving economic nuclear power in ten years. The following year the AEC announced they would assist industrial projects by waiving all charges for the use of fissionable material and would subsidize research and development needed for the construction of nuclear power reactors.

In late 1957, the first nuclear power reactor to be connected to an electrical distribution network started operating at Shippingport, Pennsylvania. Heavily subsidized by the Atomic Energy Commission, this reactor was built to demonstrate the commercial feasibility of nuclear power. Also in 1957, General Electric began construction of a large nuclear power reactor for Commonwealth Edison at Dresden, Illinois; completed in the remarkably short time of about four years, this was the first privately financed nuclear power reactor to go into operation. That same year, Babcock and Wilcox started construction of the Indian Point, New York nuclear power reactor for Consolidated Edison. By the close of 1965, five operating nuclear power reactors were "on line."

To get started in the power reactor business, the early nuclear power plants were *turnkey* plants—that is, the reactor manufacturer agreed to deliver a complete operating power plant at a fixed price, subject only to modification in case of inflation. Long-term commitments were often made for other factors affecting the cost of electricity, even including an agreement to supply fuel elements at a predetermined price. This turned out to be a financial disaster for companies such as Westinghouse when uranium prices went through the ceiling some years later. In 1964, General Electric published a price list for their nuclear power plants, indicating for example that the million-kilowatt "large economy model" could be had for some $100 million.

By the end of the 1960s, optimism was at its peak and orders for nuclear power plants were rolling in. Both government and industry were forecasting a rapid growth of cheap and reliable nuclear power. In 1966, Alvin Weinberg, director of Oak Ridge National Laboratory, told the National Academy of Science that nuclear reactors appeared to represent the cheapest source of power. Reactor manufacturers vied with each other to see who could develop the most powerful system—with a corresponding decrease in the *estimated* cost of electricity.

The problem was that nobody had any really good idea of the true cost of nuclear power. Reactor manufacturers had absorbed large losses from their original turnkey projects, and after about 1965 they no longer gave firm price guarantees. Nevertheless, in view of the widespread unbounded enthusiasm, the electric utilities simply accepted on faith the prediction that nuclear power would be cheaper than any other form of power. (Note that by this time it was taking some eight years to complete a power reactor project, and it would be several more years before accurate costs could be obtained.) By the late 1960s, it was obvious that nuclear power reactors were far more complicated and far more expensive than had been generally assumed, and cost overruns were showing up everywhere. These were casually attributed to a "prelearning" experience for a new technology, and it was taken for granted that future costs would decline as problems were worked out.

During this period of rapid growth for nuclear power, the Atomic Energy Commission had continued to play a dual role. On the one hand, their mission was to encourage and promote the widest use of nuclear energy by our free enterprise system. On the other hand, they were to "regulate" all uses of nuclear energy. This often led to compromises, particularly in the area of safety. Practically everyone with any background in nuclear energy was either directly or indirectly dependent on the Atomic Energy Commission—and understandably reluctant to bite the hand that fed them. Nuclear critics were few and far between and could be easily refuted by an army of government experts.

By the 1970s, however, critics of nuclear power were becoming more vocal, and it was obvious that the general public was concerned about many of the aspects of nuclear energy. In 1974, the Energy Reorganization Act abolished the Atomic Energy Commission and transferred its licensing and regulatory functions to the newly established Nuclear Regulatory Commission. The mission of this new agency was ". . . to see that the civilian use of nuclear materials and facilities are conducted in a manner consistent with the public health and safety. . . main focus on the use of nuclear energy to generate electric power. . . ." Other functions of the Atomic Energy Commission, such as the promotion of nuclear power, were transferred to the new Department of Energy. Military questions were to be left entirely up to the Department of Defense.

The Nuclear Chain Reactor

To understand present doubts concerning nuclear power, it may help to go over the basic principles involved in the utilization of this form of energy. The individual chemical elements that make up our universe all exist in the form of atoms, the *atom* being the smallest unit that still maintains the properties of a particular element. Two or more atoms of the same or of different elements may combine to form a *molecule.* Each atom is made up of two parts: a central core, the *nucleus,* is made up of protons and neutrons and contains most of the mass of the atom, while negatively-charged electrons revolve around the nucleus much like a small planetary system. In some instances, the atoms of a particular element will all be exactly alike; in others, the atoms of a particular element may have different weights or masses (measured in *atomic mass units* or amu). Atoms of the same element that differ in weight are known as *isotopes* and are normally defined in terms of their individual mass units. For example, natural uranium as it comes from the ground consists of two isotopes, uranium 238 (which makes up about 99.3 percent of the element) and uranium 235 (about 0.7 percent).

Another element which has been important in the development of nuclear energy is hydrogen; this consists of two isotopes, hydrogen 1 (ordinary hydrogen), which makes up most of the element, and hydrogen 2 (deuterium), which is present to about 0.015 percent. If deuterium is substituted for ordinary hydrogen in water, the compound is known as *deuterium oxide,* or *heavy water.*

In an ordinary *chemical* reaction, such as when hydrogen is burned to form water, two atoms of hydrogen combine with one atom of oxygen to form one molecule of water. This involves a rearrangement and sharing of the planetary electrons of the atoms, but there is no change in the nucleus and both isotopes of hydrogen react in exactly the same way.

In *nuclear* reactions, such as those involved in nuclear power reactors, there is a change in the nucleus of the individual atoms that react. The various isotopes react entirely differently in the case of nuclear reactions; furthermore, by tapping the energy of the nucleus, about a million times as much energy is produced as in the case of a chemical reaction. In addition, an unwanted by-product of nuclear reactions results from the fact that *fission,* or splitting of the nucleus of an atom produces a different chemical element—usually one that is radioactive.

The present source of nuclear energy is the isotope uranium 235, which can capture a neutron according to the following reaction:

uranium 235 + neutron → fission products + 2.5 neutrons + energy

Actually, the fission of a uranium 235 nucleus can occur in many different ways, but on the average two and one-half new neutrons are produced for each fission. Since the fission reaction produces more neutrons than it consumes, a self-sustaining chain reaction is possible if at least one of the product neutrons can be captured by another uranium 235 nucleus to continue the fission process.

A first glance, it might seem that the design of a nuclear chain reactor would be a relatively simple task, since the fission reaction produces a substantial excess of neutrons. Natural uranium, however, contains only 0.7 percent uranium 235, and the remaining 99.3 percent (uranium 238) does not contribute to the fission process. Also, the fission reaction produces fast-moving neutrons which can be rather easily captured by the abundant uranium 238 isotope before they get a chance to continue the nuclear chain reaction. It is thus impossible to establish a chain reaction in a system of pure natural uranium —a very fortunate fact which prevents use of pure natural uranium for the construction of nuclear weapons.

If uranium 238 is causing problems, why not just get rid of it? This has turned out to be a lot easier to say than to do, and the separation of isotopes is a very difficult and expensive job. During World War II, several different isotope separation processes were investigated simultaneously without assurance that any would work. The *gaseous diffusion* process proved successful, however, and a large plant was constructed at Oak Ridge, Tennessee. In this process, uranium hexafluoride gas (a compound of natural uranium) is passed through a porous barrier several thousand times to effect a substantial overall separation into a product with over 90 percent uranium 235.

Use of a Moderator

In addition to removing undesirable uranium 238, there is another way in which natural uranium can be used to establish a nuclear chain reaction. It can be used with a material known as a *moderator,* which effectively slows down the fast-moving neutrons so that the resulting slow neutrons can be used more efficiently by the 0.7 percent uranium 235 present, thus making it possible to create a self-sustaining chain reaction.

If natural uranium is used as fuel in a nuclear chain reactor, the choice of moderator is limited pretty much to two materials, graphite or heavy water. Both graphite/natural-uranium and heavy-water/natural-uranium reactors were developed by the United States during World War II. Graphite has a rather substantial advantage in that it is inexpensive and can be easily made from petroleum coke; the graphite/natural-uranium reactor, however, must be quite large, about a 25-foot cube, and its power output is limited. On the other hand, heavy water is a much more efficient moderator, and the size of the system can be reduced to 6 or 8 feet; its disadvantages are its very high cost and the difficulty of producing it in the required amounts.

The graphite/natural-uranium reactor has been selected by the United Kingdom as the basis for much of its nuclear power program. With limited reserves of high-cost coal and still strongly dependent on imported fuels at present, they have considered development of nuclear energy as vital to the national interest. The power reactor developed by the British consists of a large cube of graphite with fuel channels containing uranium rods that are jacketed with an aluminum-magnesium alloy. Carbon dioxide gas circulates over the fuel rods and passes to a conventional steam generator. Advanced high-temperature reactors are also under development; these will use helium as a coolant and will provide very efficient conversion of heat into work.

The heavy-water/natural-uranium power reactor has been widely studied in Canada and is being developed as the basis for their nuclear power program. Since the Canadians have plenty of hydropower, there is much less urgency for them to develop nuclear power; however, in view of their ample supplies of domestic uranium and with an improved process for the production of heavy water, nuclear power is regarded as an eventual source of energy when needed. (Incidentally, the United Kingdom has also recently begun to show some interest in the development of a heavy-water power reactor.)

The United States has selected a reactor system moderated by ordinary (*light*) water for its nuclear power program, and this type of system is also being widely used in Germany, Japan, and Sweden. Since light water is both an excellent coolant and a very good moderator, an extremely high power output can be achieved from this type of system. It does have one real disadvantage in that natural uranium cannot be used. Light-water reactors require a more expensive *partially enriched* uranium—that is, it must first be concentrated from 0.7 to about 3.5 percent uranium 235 in a gaseous diffusion plant.

The power reactors used in the United States at this time all consist of vertical *fuel elements* (small pellets of uranium dioxide enclosed within *cladding* tubes of stainless steel or zirconium, a rare metal). The fuel elements are stacked together in a grid pattern within a very thick-walled *reactor pressure vessel,* built to withstand some 2500 pounds per square inch. Ordinary water at high temperature and pressure passes down around the outside of the fuel tubes to remove heat. The entire reactor vessel and its auxiliary equipment are located inside a large *containment building,* designed to prevent leakage of radioactive material in the event of an accident.

In the *pressurized-water reactor* (PWR) as marketed by Westinghouse, Combustion Engineering, and Babcock and Wilcox, there is no boiling water in the core of the reactor pressure vessel; the coolant water passes to a steam generator which produces steam for a turbine. In the *boiling-water reactor* (BWR) as marketed by General Electric, boiling does occur in the reactor core, with the steam (in this case, slightly radioactive) passing directly to the turbine. Modern nuclear power reactors have quite high generating capacities, ranging from 1 million kilowatts and up. They are also extremely expensive to build, as we mentioned before, with the costs now running into billions of dollars.

The Question of Safety

In discussing the safety of nuclear power reactors, let us first make one important point: While it is not impossible for a nuclear power reactor to overheat and blow up, just as a conventional fossil-fired boiler can overheat and blow up, *there is no way a nuclear power reactor can turn into an atomic bomb.* A bomb, which is the only true nuclear explosion, requires the assembly of essentially pure uranium 235 or plutonium 239 in a time interval of 1 millionth of a second or less. A conventional nuclear power reactor contains only low-enrichment uranium, there is a lot of light-water moderator present, and it would be physically impossible to move a large power reactor assembly in a fraction of a second. It is completely false to refer to nuclear power reactors as "incipient atom bombs."

Nevertheless, a nuclear power reactor is potentially more dangerous than a conventional steam boiler. When fossil fuels are burned, there is a certain maximum temperature and maximum heat production that are both determined by the thermodynamics of the particular reaction. In a nuclear reactor, the temperature and heat production both depend on the intensity of the nuclear chain reaction which, theoretically, has no upper limit. Once a nuclear reactor has become *critical* (that is, operating to produce power at a constant rate), the nuclear chain reaction can rise to dangerously high level within a few seconds. The primary danger would be in the release of large quantities of highly radioactive fission products rather than the explosion itself.

The nuclear chain reaction is dependent on a balance between production and loss of neutrons. Nuclear power reactors are controlled by so-called "poison" *control rods,* which instantaneously absorb neutrons. When fully inserted into the reactor core, the control rods gobble up so many neutrons that a self-sustaining chain reaction becomes impossible. By slowly withdrawing the control rods, the chain reaction can be gradually adjusted until there is an exact balance between the production and loss of neutrons—in other words, when the system is critical. In practice, several other methods may be used to control power reactors—for example, the pressurized-water reactor is also controlled by dissolving boric acid, a soluble neutron absorber or "poison," in the reactor coolant water; the boiling-water reactor may also be controlled by adjusting the recirculation rate of the coolant to the reactor core, thereby changing the amount of steam in the core.

The basic hazard, as we have pointed out, is the radioactive fission products which are produced within the fuel elements. Every fission of uranium 235 produces two fission products, and almost all of these are radioactive. This is complicated by the fact that fission occurs in many different ways, so that the fission products consist of a mixture of many different radioactive chemical elements with radioactive *half-lives* (the time required for the radioactivity to fall to half its original level) ranging from a few seconds to millions of years. It is difficult to chemically separate and concentrate this mixture of radioactive materials since chemical methods that work with one element may not work with another. Furthermore, the composition of the fission products is continually changing with time, so that one particular radioactive element may be prominent today, another next week, and still another next year.

Under normal operating conditions, these radioactive fission products are confined within the fuel elements and cannot escape unless there is a failure of the stainless steel or zirconium cladding material. Radioactive fission products continue to produce very large quantities of heat even after a nuclear power reactor is shut down. For this reason, a dependable form of auxiliary emergency cooling is required to take over in the event of interruption of the regular coolant flow. If cooling is stopped for even a short period of time, the *radioactive decay heat* may vaporize water from the core and the exposed fuel elements will then overheat, rupture the cladding, and under severe conditions may even melt. Radioactive fission products can then pour into the primary coolant and be released from the reactor pressure vessel into the containment building. *This is exactly what happened at the ill-fated Three Mile Island reactor in the spring of 1979.*

Why is radioactivity such a problem in the case of nuclear power reactors? Actually, it has been around a long time—radioactivity was discovered in 1896 by the French physicist Henri Becquerel, and we have had some 85 years' experience in the handling of radioactive materials. The problem arises because of the tremendous amounts of radioactive fission products produced by a nuclear reactor. For example, at one time 1 single gram of radium was considered to be a large quantity of radioactive material; a modern nuclear power plant produces some 4000 grams of radioactive fission products each day—more than 1 ton over the course of a year. This mind-boggling quantity is so much larger than anything encountered previously that it is small wonder there are doubts about the feasibility of handling and storing the material.

Risk Evaluation

Today's considerable controversy over the safety and reliability of nuclear power is undoubtedly the direct result of the excessive optimism expressed by government agencies, reactor manufacturers, and public utilities during the years that preceded the incident at Three Mile Island. For years the Atomic Energy Commission had a monopoly on nuclear power—they were quite literally judge, jury, and executioner. Their primary mission was to *develop* nuclear power and, understandably, they tended to overlook negative factors. Reactor manufacturers were only too glad to climb on the bandwagon, citing confident and optimistic government figures. Electric utilities were in no position to question the information passed on to them by these experts.

On the few occasions when doubts were raised as to the safety of nuclear power, these were simply brushed aside with the assurance that the nuclear experts had everything under complete control. Under such conditions, and in view of the fact that electric utilities in general have never been very frank in dealing with the public, it was no wonder that serious questions began to arise, together with a strong distrust of statements made by nuclear advocates. The nuclear critics gradually developed experts of their own, who forcefully expressed the most negative aspects of nuclear power. Today, nuclear advocates and nuclear critics have become so polarized in their views that there appears to be no possibility for any reasonable evaluation, let alone a chance for the plain facts to be heard by a very confused public.

Two acknowledged risks must be considered in the use of nuclear power. First, the tremendous quantities of radioactive fission products that are produced in a nuclear power reactor could, in the event of a serious reactor accident, escape to the atmosphere and contaminate large areas near the plant. Under the very worst conditions, hundreds or even thousands of people could be killed, and many square miles of land would have to be evacuated for a long period of time. The probability of such an accident is very, very small. Nuclear power reactors are provided with elaborate safety controls; in addition, the entire reactor system is enclosed in a large reinforced-concrete containment building designed specifically to prevent the release of any radioactive materials. The potential risk involved here depends on a very small probability applied to a very substantial accident.

The second risk results from the fact that nuclear power plants under normal operating conditions do release a certain amount of additional radiation to the atmosphere. Even without the presence of nuclear reactors, we are all continually subjected to radiation both night and day from cosmic rays and various radioactive materials present in the environment and in our bodies. We know, for example, that small amounts of radioactive carbon and radioactive potassium, as well as varying amounts of other radioisotopes such as radium, have always been ingested with our food and water. While there is nothing we can do about this *background radiation,* we also very casually accept rather substantial additional exposure to radiation from medical X rays, dental X rays, jet travel, and the like. The fact is that a nuclear power reactor will cause much less radiation exposure to people living in the vicinity of the plant over a period of a year than their natural background and *much less* than would be encountered in a dental or chest X ray. While many radiobiologists believe this small additional exposure to be completely meaningless, others believe that *any* additional exposure is harmful and may cause a small increase in the future incidence of such diseases as leukemia and bone cancer. Here, again, it is impossible to qualify the degree of risk—or in this case to determine whether there is any risk at all.

We must also keep in mind that any generation of electricity involves some risk. Coal mining, for example, has always been one of the most hazardous occupations in the United States. Each year there are some 200 deaths, 25,000 lost-time injuries, and tens of thousands of miners who apply for compensation because of "black lung" disease. Most coal today is used for the generation of electricity, so it is rather obvious that our coal-fired power plants entail a substantial risk. In addition, we have seen that these plants emit large quantities of solid particles and noxious gases that may cause deaths from asthma and other respiratory diseases, the production of carbon dioxide increases the probability of atmospheric changes through the "greenhouse" effect, and we are only just beginning to comprehend the harmful effects of acid rain—the general public clearly "pays a price" for its coal-fired power plants.

There have been numerous studies made of the comparative risks of electricity generated by various means. All of these have involved arbitrary and highly inaccurate assumptions. There is general agreement, however, that nuclear power involves both a much smaller occupational risk and a much smaller risk to the general public than power from coal, oil, or natural gas. Nuclear critics disagree with these studies and feel that we have not had enough experience to accurately evaluate the hazards of nuclear power—once more we are at an impasse.

Have there been any serious accidents associated with commercial nuclear power reactors in this country? The most serious was the well-publicized Three Mile Island accident, in which many people living in the vicinity of the reactor did receive some additional radiation exposure. The amount was very small, equivalent perhaps to what would be encountered in a dental X ray. While there is no real evidence of physiological harm to those living near Three Mile Island except possibly a slightly higher probability for cancer in the next twenty or thirty years, the psychological effects have been disastrous. Thoroughly confused by the on-again, off-again evacuations and by conflicting statements from government and industry experts, these people (and much of the general public) were only too ready to accept the implication that a real catastrophe had occurred.

Radioactive Wastes

Two kinds of radioactive wastes are encountered in the operation of a nuclear power plant. *Low-level wastes* are those to be found in cooling water, gaseous effluents, contaminated clothing, radioactive tools, and the like. They are not particularly hazardous. Radioactive water is often simply pumped back into the regular reactor cooling water system, radioactive gases are filtered and discharged to the plant stack, and solid wastes are compacted and sent to a disposal area where they are buried and covered with earth.

High-level wastes are those from the radioactive fuel elements and are a far more serious matter. Under normal operation, fuel elements remain in a power reactor for three years, during which time some 40 to 50 percent of the uranium 235 is consumed by the fission process. When discharged from a reactor, the spent fuel elements still contain one-half or more of the uranium 235 originally present, plus tremendous quantities of radioactive fission products, radioactive materials produced from neutron irradiation of the cladding material, and higher uranium isotopes such as uranium 236 (produced by neutron capture in uranium 235) which do not lead to fission. The irradiated fuel elements also contain over 1 ton of toxic plutonium.

When the nuclear power program in the United States was first started, it was assumed that spent fuel elements from the commercial power reactors would be chemically processed to remove the radioactive fission products and thereby recover pure uranium (still containing about 2 percent uranium 235) and plutonium. The uranium was to be recycled back to a gaseous diffusion plant for enrichment and reused in the nuclear power reactors. The plutonium was to be taken as a "credit" and eventually used as fuel for advanced reactors then in the planning stages. At that time, the Atomic Energy Commission's processing facilities at Richland, Washington, and at Savannah River, South Carolina, were producing plutonium for the weapons program, and a third facility at the National Reactor Testing Station in Idaho Falls, Idaho, was recovering highly enriched uranium from the large test reactors. It was assumed that private industry would construct independent fuel processing facilities which would become available to the nuclear power industry on a contract basis.

Unfortunately, the Atomic Energy Commission was so involved in promoting nuclear power at this time that they underestimated the difficulties of reprocessing and put it on the back burner—after all, it would take years to build a nuclear power plant and still more years of operation before large numbers of spent fuel elements became available. They finally got around to the reprocessing question in the 1960s. Nuclear Fuel Services, Inc., a joint undertaking of W.R. Grace and American Machine and Foundry, was issued a license for a reprocessing plant at West Valley, New York. This plant was operated from 1966 to 1972 and was the first such privately owned facility in the world. There were continuous problems with radioactive effluents and with excessive radiation exposure of employees. In 1972, under new ownership, the plant was closed for renovation and enlargement; because of the extremely high cost of meeting new and strict environmental requirements, it was never reopened. During six years of operation, it had accumulated some 600,000 gallons of highly radioactive wastes that were stored "temporarily" in carbon-steel tanks; *these wastes are still there,* and it is now estimated that ten to fifteen years and hundreds of millions of dollars will be required to process and ultimately dispose of them.

Also in the late 1960s, General Electric built a facility at Morris, Illinois, which was intended to use a variation of a process used to recover weapons-grade plutonium from nuclear wastes. The process was a technical failure, and the plant never got beyond the test stage.

The Problem of Plutonium

Meanwhile, another problem had arisen which has turned out to be far more serious than anyone anticipated. All power reactors contain substantial quantities of uranium 238, which can capture neutrons according to the following reactions:

uranium 238 + neutron → uranium 239 → neptunium 239 → plutonium 239

When uranium 238 captures a neutron, it forms radioactive uranium 239, which changes in a few days to long-lived plutonium 239. Like uranium 235, plutonium 239 can be used as fuel for a nuclear power reactor; unfortunately, it turns out that plutonium 239 is even better than uranium 235 for the production of atom bombs. Each week, a large power reactor produces enough plutonium for perhaps one atom bomb—or some fifty bombs per year. True, the plutonium produced in power reactors is not the best quality for bomb purposes because of the presence of higher isotopes such as plutonium 240*, but it can be used to produce an atom bomb thousands of times as destructive as the standard-type bombs used in World War II over Germany. The point here is that *any* reprocessing of fuel elements from nuclear power reactors would involve much routine handling and accounting for thousands of kilograms per year of plutonium. It is not hard to imagine how small amounts could be successfully "diverted," leading to possible widespread proliferation of nuclear weapons.

The Waste Problem Today

Even after the failure of the private processing plants, the federal government implied that adequate facilities would be made available to the utilities. Originally, it had been planned to store spent fuel elements for one to three years under water in storage pools at the site of each power reactor; at the end of this time, the fuel elements would be shipped for reprocessing. In 1976, President Ford warned America's nuclear power industry that further fuel reprocessing might be unacceptable because of the danger of proliferation. The following year, President Carter announced an outright ban on fuel reprocessing, to be in effect indefinitely. Suddenly these huge plants had no prospect of

*Plutonium 240 has a high spontaneous fission rate that produces an undesirable neutron background. Weapons-grade plutonium is produced by using short fuel cycles that minimize higher isotopes.

ever getting rid of their spent fuel elements! Most of them have now been forced to expand the capacity of their fuel storage pools and, in some cases, to construct new holding facilities. Many of these will be completely full by the mid-1980s, forcing the utilities to send fuel elements elsewhere, build new storage facilities, or simply shut down their plants.

Proposals to build large away-from-reactor pools for regional or national storage of spent fuel elements merely transfer the problem from one place to another and represent no real solution. There is absolutely no question that the federal government will have to face this serious radioactive waste disposal problem. The Department of Energy is currently looking into the possibility of establishing a national storage site for radioactive wastes in some "stable" geological formation. One suggestion is to store wastes in a salt cavern several hundred feet below the surface of a remote location in southeastern New Mexico. Some experts feel that granite, basalt, or shale might be preferable. (Still others have actually proposed to pump the wastes deep within the earth, to deposit them in the ocean bed, to store them on polar ice sheets, or even to shoot them into outer space!)

If nuclear power is to remain a viable source of energy in the United States, the dreaming will have to stop; hard-nosed decisions will have to be made—and soon.

The Fast-Breeder Reactor

We have seen that nuclear fuel cycles used today in the United States start with natural uranium containing only 0.7 percent uranium 235. This is first run through a gaseous diffusion plant to produce a partially enriched uranium which contains roughly 3.5 percent uranium 235 and a waste product which contains roughly 0.3 percent uranium 235. The partially enriched uranium is then used for reactor fuel elements, with perhaps half of the uranium 235 being consumed during the three-year lifetime of the fuel element. Only some 0.3 percent of the natural uranium is actually used to produce power, and 99.7 percent is essentially wasted.

It is not surprising that many nuclear experts consider present-day reactors to be wasteful and inefficient. The *fast-breeder reactor* (FBR) offers a means of using *all* the natural uranium, including the 99.3 percent uranium 238 that is presently wasted. In comparison with today's technology, the fast breeder would increase our utilization of nuclear fuel by several hundred times and thereby provide fuel reserves equivalent to thousands of years at today's rate of energy consumption.

The fast-breeder reactor uses plutonium 239 as fuel according to the following reaction:

plutonium 239 + fast neutron → fission products + 3 fast neutrons

The fast breeder would consist of a central fuel region containing the plutonium 239 surrounded by a *breeding blanket* which contains natural or depleted uranium (chiefly uranium 238). Fast neutrons from the core would leak into the blanket region where they would be captured by uranium 238 to produce plutonium. The success of the fast breeder depends entirely on the neutron balance; since fast fission of plutonium produces on the average three new neutrons, one can be used to continue the chain reaction, one can be used by uranium 238 to replace the atom of plutonium 239 that was used up by fission, and one is available to supply losses and possibly to further increase the amount of plutonium produced. Fast breeders now under development are expected to produce about 20 percent more plutonium than they consume.

Despite obvious advantages, the fast breeder presents many serious problems. Hundreds of kilograms of plutonium are required just for the initial fueling of the reactor core. Both the breeding blanket and the spent fuel elements must be chemically processed to purify and recover the plutonium, thereby opening up another possibility for proliferation of atomic weapons. Since conventional coolants such as water cannot be used in a fast breeder, it is necessary to resort to pure liquid sodium, a substance that is very corrosive and burns instantly on exposure to air or water. Fast-breeder reactors are also more difficult to control, requiring a more sophisticated control system, which naturally increases the possibility of an accident that might release large quantities of plutonium.

A prototype fast breeder is presently under construction by the federal government on the Clinch River about 30 miles west of Knoxville, Tennessee. It is designed to produce 350,000 kilowatts of electricity and is expected to serve as the basis for future full-scale fast breeders that will produce over 1 million kilowatts of electric power. Because of his real concern over possible proliferation, President Carter tried to cancel the Clinch River breeder project; he was overruled by Congress. In addition to work in the United States, the technology of fast-breeder reactors is now being far more actively pursued by France, England, and West Germany.

There appears to be no question that we may have to eventually go to the fast breeder for a source of electricity. It will probably be at least ten years before we obtain operating experience from the Clinch River project and another ten years before relatively safe large-scale fast breeders can be designed and constructed. The fast-breeder type of reactor will certainly not be much help to us for the balance of the twentieth century.

The Future of Nuclear Power

Except for nuclear power reactors already in operation or under construction, our nuclear power industry is now at a complete standstill. Many factors have contributed to this. Because of declining growth rates, the utilities do not foresee the need for more large generating capacities, and this has already lead to cancellation of many reactor projects. Since power reactors have become almost prohibitively expensive, particularly with the many safety controls now required, coal-fired plants are looking better from an economic standpoint. Licensing and regulation procedures for a power reactor are now so drawn out and complicated that it requires some *twelve* years to bring a reactor project to fruition, in comparison to perhaps half that time for a coal-fired plant. In addition, there is now growing concern as to the future availability and cost of nuclear fuel—and the situation is not helped by the unsolved problem of radioactive waste disposal.

What about the safety question? The Three Mile Island incident has indeed dealt a body blow to the nuclear power industry, but are power reactors safer now? The Nuclear Regulatory Commission has instituted additional safety standards and strict training requirements for the operators of power reactors. In some cases, they have required modification of operating reactors as well. One problem is that, after a certain point is reached, safety becomes almost self-defeating as more and more sophisticated controls are added; each new gadget is just one more thing that can go wrong. It is also possible to make a power reactor so complicated that it can no longer be operated effectively—the slightest perturbation causes it to automatically shut down!

As far as the United States is concerned, it appears that nuclear power will continue at about its present level for the rest of this century. This is perhaps not a bad idea since it will provide us with time to reassess our situation and to develop more experience with power reactor operation, particularly with the fast breeder. On the other hand, this does not apply in most of the rest of the world. Japan and western Europe are currently far more dependent on imported fuels; they need nuclear power now and hence can be expected to compromise when it comes to such questions as safety and proliferation.

How real is the proliferation threat? Since any nuclear reactor operating at an appreciable power level can produce plutonium, the only sure way to stop the spread of atomic weapons is to stop the construction of all reactors. Several years ago, for example, India was provided with an experimental nuclear reactor for "peaceful" research and training. The United States supplied the heavy water, and Canada supplied the uranium fuel. The Indians proceeded to recover plutonium from the reactor fuel elements, and they were not long in producing their first atom bomb.

There are two considerations that might lead us back to nuclear power sooner than seems likely now. First, our oil and our coal have become essential for other purposes such as the production of liquid fuels and as raw materials for our entire chemical industry. It is extremely wasteful to burn these as fuels for the generation of electric power. Since nuclear fuels have no value other than as a source of power, it may well become necessary to use uranium for power and reserve our remaining fossil fuels for these other needs.

Second, it should be pointed out that the nuclear fuels have one very substantial advantage in contrast to the fossil fuels: *they do not produce any carbon dioxide.* As will be discussed in Chapter 6, our increased use of the fossil fuels has caused a gradual increase in the carbon dioxide content of the atmosphere. There is now sufficient evidence to show that this could lead to very serious worldwide environmental changes, serious enough to place a limit on any future growth in the use of conventional fuels. If we can solve the problem of radioactive waste disposal, it is entirely possible that nuclear power may some day be more acceptable environmentally than power from fossil fuels.

REGULATION OF THE ELECTRIC POWER INDUSTRY

Since an electric utility is obviously a monopoly, the need for regulation was recognized from the earliest days of the industry. The first power stations were quite small, so that initial regulation was entirely local and probably not very effective. By the start of the twentieth century, however, various states had begun to exercise control over the utilities through the establishment of public utility commissions to regulate rates and profits. During the 1920s, there was a very rapid growth of electric utilities beyond state lines; the utilities became so large and so complex that it was difficult for the states to control them. To build larger systems and thereby pyramid their profits, many of the utilities were taken over by *holding companies* with names such as Electric Bond and Share. The fact was that in most cases the holding companies were formed primarily to get around state laws.

The Insull Story

One of the most interesting and most publicized examples of such holding companies was the utility empire built up by Samuel Insull in the Chicago area. Born in London, England, in 1859, Insull helped to organize the telephone business in England. In 1881, this bright young man came to the United States under a direct invitation from Thomas Edison, and he was soon in charge of Edison's personal and business affairs. In 1892, Insull went to Chicago to become president of the Chicago Edison Company; six years later, he consolidated this company with several smaller electric utilities to form the Commonwealth Edison Company, with himself as president. By 1907, the electric power production of Commonwealth Edison exceeded that of New York, Brooklyn, and Boston combined.

In 1912, Insull organized the Middle West Utilities Company as a large holding company to service a "superpower" electrical system covering much of the Midwest. In 1919, he became president of the People's Gas Company of Chicago, and during 1929–31 he was instrumental in bringing natural gas from Texas to Chicago and surrounding cities. Many of the concepts Insull developed became standard utility practice. By 1930, his holding companies had assets of over $2 billion and were producing a tenth of the nation's electric power.

Although Insull securities weathered the 1929 stock-market crash, a further drop of the market in 1931 caused his empire to collapse and sent many of his companies into receivership. Insull became a focus of public bitterness in the hard days of the depression and found himself a political issue in the 1932 elections. Foreseeing only persecution, he fled to Greece. He was returned dramatically by the United States government in 1934 to face charges of mail fraud, embezzlement, and violation of the bankruptcy acts, and, at the age of 74, was sent to jail for a time because he was unable to raise the required $200,000 bail. After three much-publicized trials, however, he was cleared of all charges and allowed to spend the remaining four years of his life in relative peace.

Federal Regulatory Agencies

The collapse of holding companies during the depression of the 1930s gave the utilities a bad name; many states increased their regulatory activities, and the federal government actually made motions to nationalize the entire electric utility industry. President Franklin Roosevelt received much support when he advocated public operation as a "birch rod" to control the corrupt private systems. In 1935, the Federal Power Commission was empowered to regulate interstate wholesale rates for electricity, and in the same year the Securities and Exchange Commission was authorized to deal with the financial and administrative structure of many large electric utilities. Holding companies were now essentially limited to a single integrated utility system, and service organizations were required to serve operating companies at cost.

Following World War II, there was another period of very rapid growth for the electric utility industry. The production of electricity in the United States more than doubled between 1945 and 1955; the next ten years saw another doubling, and by the 1970s it had doubled once again. Smiling utility heads were planning future expansion based on an assumed growth of some 7 percent every year!

At the same time, there was a trend toward much larger generating facilities. In 1945, the privately owned utilities were operating some 2000 plants, most of them producing less than 100,000 kilowatts. Thirty years later, the number of plants had dropped slightly, but over half the production of electricity was from plants greater than half a million kilowatts. Some seventy plants were greater than 1 million kilowatts—including approximately twenty large nuclear plants.

During this boom period of the 1950s and 1960s, the electric utilities were enjoying a gradual decrease in the cost of producing electricity. The larger units produced electricity at a lower cost per kilowatt than the smaller units; in addition, technological improvements gradually increased the efficiency for the conversion of heat into work, thereby reducing fuel requirements.

In these carefree years, the electric utilities had a "sweetheart" relationship with the regulatory agencies. State public utility commissioners were very often political appointees—or even old cronies of the governor. Consumer groups and politicians were content as long as the price of electricity was not going up. Regulatory agencies had the responsibility for determining a "fair" rate of return to the utility, nominally based on the plant investment, but there was considerable latitude in defining what was actually fair. Most state regulatory agencies were only too glad to "live and let live" as long as there were no complaints. Utilities that increased their rates of return without raising prices were permitted to keep any additional money they brought in. No questions were asked if the rate of return seemed to increase above an "allowed" level.

The End of the Honeymoon

This idyllic state ended in the late 1960s when the industry was hit simultaneously with higher inflation, soaring interest rates, and increased equipment costs, particularly for nuclear power systems. The utilities had to persuade state regulatory agencies to approve higher electric rates, but this involved time-consuming regulatory reviews. Costs were increasing so fast that before one rate increase could go into effect it was necessary to file for still higher rates. Furthermore, consumer groups became active and the utilities found themselves faced with the need of opening their books and defending themselves to an increasingly critical public.

The next major blow to the utility industry was the Arab embargo of 1973–74, which led to a quadrupling of the price of OPEC oil. This immediately affected the price of the some 17 percent of United States electric power that came from the burning of petroleum products. Also, higher prices for crude oil immediately triggered higher prices for all other fuels. In most cases, the utilities were able to pass these increased costs on to their customers via the "fuel adjustment" clause. As far as the average consumer was concerned, this just meant paying more for electricity.

During the embargo, the government asked for conservation of all energy:

thermostats were turned down, outdoor lighting was reduced, and customers tried to reduce their use of electricity as much as possible. Total sales of electricity decreased in 1974—the first yearly reduction since the depression of 1929–33. While sales bounced back in 1975, yearly growth is now averaging only some 3.5 percent, half the "traditional" growth previous to the embargo.

On April 23, 1974, Consolidated Edison of New York announced it was suspending payment of its quarterly dividend. Furthermore, unless New York State was willing to pay $500 million for two partially finished generators, there was considerable doubt the company could meet future electric demands! Long considered by investment counselers as a safe investment for "widows and orphans," Consolidated Edison had paid uninterrupted quarterly dividends for eighty-eight years. The potential collapse of this giant utility sent shock waves throughout the entire industry, and worried investors began to wonder who would be next.

Today there remains considerable question as to the financial stability of the United States electric utility industry. Lower profits and higher interest rates have made it increasingly difficult to obtain funding for the very substantial costs of new generating facilities. Fuel prices are increasing even faster than inflation in general. The thermodynamic efficiency of electric generation has reached a peak unless an entirely new technology can be developed (in fact, it has even decreased in recent years because of the somewhat lower thermodynamic efficiency of nuclear plants). Increasingly militant consumer groups are opposing rate increases—often successfully—and demanding that the regulatory agencies revise the entire rate structure and consider social welfare problems in determining the new rates. Environmentalists are using regulatory hearings as a forum to advance detailed siting criteria for new power plants and, in some cases, to argue against proposed nuclear plants. Meanwhile, Mary and John Q. Public become more and more confused and irritated by the whole circus and wonder how to handle the ever-rising electric bills.

ENERGY ALTERNATIVES

One of the primary factors that got us into our present situation was the "cheap energy" dream that lasted for so many years—we were, in fact, victims of our own efficient technology. During the carefree postwar years, gasoline sold for less than 30 cents a gallon in most of the United States, and Americans encouraged each other to drive the powerful gas guzzlers that their automotive industry was only too happy to supply—complete with chrome trim and air conditioning. In Europe, where gasoline at that time was highly taxed and sold at $1 a gallon or more, Europeans demanded and got small, fuel-efficient cars.

We know that the most urgent problem this country faces in the next few years is the need to reduce our consumption of liquid fuels (gasoline, fuel oil, diesel oil, and jet fuel). In an economy of scarcity, there are two ways to limit consumption: higher prices to discourage use, and rationing. Since gasoline is the largest refinery product in the United States, its consumption could and should be reduced by raising the federal tax as high as is necessary to achieve results, with the added income used to develop alternate fuels (and not, as in the past, to build more highways). Rationing should be considered only as a last resort. Rationing was not very successful even during World War II when the military need was obvious and when tankers were being torpedoed off New York harbor. The recently proposed rationing scheme involving a "white market" for gasoline coupons is something only Washington could dream up. Apart from the tremendous cost of administering such a plan and the inevitable thefts and counterfeiting of coupons, the proposal to issue coupons based on the number of registered vehicles was grossly unfair and would surely have resulted in the registration of every junk car that could be held together with tape and baling wire long enough to collect some coupons.

The automotive industry can play a dominant part in reducing energy consumption by designing cars that are simple, easy, inexpensive to repair, and built to last for at least 200,000 miles. Catalytic converters can be done away with by redesigning engines to meet somewhat more reasonable emission standards, and to provide minimum mileage between 40 and 50 miles to the gallon (increased mileage will in itself substantially reduce pollution as well). Expensive yearly model changes can be prohibited, with only functional changes consistent with energy goals being permitted.

Several steps were taken to reduce oil consumption during the 1973–74 Arab embargo, but these were relaxed once the immediate crisis had passed. As a minimum, the following steps should be reinstituted on a permanent basis: (a) A national speed limit of 50 miles per hour should be enforced impartially for both trucks and passenger cars. (b) Service stations should be closed on Sundays. (c) Shopping malls and all nonessential businesses should be closed on Sundays. (d) Unnecessary lighting, including all outdoor advertising signs, should be eliminated. (e) Winter thermostats should be set no higher than 68 degrees Fahrenheit for homes and offices; air conditioning should be set no lower than 78 degrees Fahrenheit. (f) Business and industry should be further encouraged to reduce employee driving by means of carpooling and/or staggered hours. (g) Most important, mass transit should be expanded to provide workable alternatives to automobile transportation.

While the United States can substantially reduce its energy consumption without suffering any genuine decline in its standard of living, there is no question that some things will have to go and some people will be hurt. For example, that portion of the tourist industry that is now dependent on auto travel will certainly suffer, while large cities with adequate public transportation may actually benefit. The entertainment industry will suffer, with the possible exception of local events. Not long ago, practically every town had its own facility for lectures, dances, town meetings, and the like. This may gradually return as people lose a certain degree of mobility. Attendance at sports events will suffer unless they can be reached by public transportation; a football stadium 50 miles out of town may well find itself playing primarily to a television audience (in this connection, more pay-as-you-go television may provide an answer). Ski resorts that now use large amounts of energy for artificial snow-making (so that thousands of drivers can use gasoline en route to and from weekends of skiing) will of necessity disappear. Snowmobiles and trail bikes will be replaced by the excitement of cross-country skiing and hiking. Campers will make way for backyard vacations. Sunday afternoon at the shopping mall will yield to walking, gardening, or even relaxing with friends at home.

Wasteful use of energy today means that someone tomorrow will not have heat for his home or gasoline to reach his job—or perhaps even have a job. College students making their annual pilgrimage to Fort Lauderdale, Florida and senior citizens traveling around the country in packaged group tours will come to realize that tremendous quantities of gasoline are going into nonproductive use. Licenses to drive may be limited to those with proven need. School busing may have to be cut back or even eliminated in most communities, with smaller neighborhood schools returning to favor. Many persons feel even now that the huge sums of federal tax money presently poured into busing should instead go into quality teaching for all our children. It is simply another example of the choices that will face us, choices that are bound to upset many of us regardless of their final outcome.

As a matter of fact, there are already two indications that the conservation message may be starting to have some impact on our energy habits. First, industry in the United States has been able to substantially reduce its waste of energy and thereby has increased the efficiency of manufacturing operations. Since the 1973–74 Arab embargo that brought "cheap energy" to an end, our energy consumption per dollar of gross national product has actually decreased by about 10 percent. Second, the total consumption of petroleum in the United States has shown a modest decrease in the last few years. Unfortunately, it is very doubtful that conservation has had much to do with the decrease in petroleum consumption; the real credit here must go to higher costs for gasoline and fuel oil and to the current business recession.

SUBSTITUTES FOR TODAY'S LIQUID FUELS

It is now obvious that the United States must do everything possible to develop alternate sources of liquid fuels. To gain any real perspective on this challenge, we must first face the magnitude of the problem. The total quantity of petroleum liquids consumed yearly in this country is about a billion tons. This is greater than the entire production of coal in this country and only slightly less than the entire world production of food. If all the coal mined annually in the United States were converted to liquid fuels, we would still have less than our present consumption! If the entire annual production of food in the world was converted to alcohol, American motorists would not have enough to satisfy their current driving "needs." All energy experts are sooner or later led to the same conclusion, that the only realistic solution to our liquid fuels problem is conservation.

A brief look at some of the possible solutions proposed to augment our supply of petroleum will show that there are very serious problems associated with the development of each of them; in addition, they will all be much more expensive than today's gasoline or fuel oil, and it is very unlikely that any of them can be developed soon enough to make a substantial contribution to our energy needs before the end of this century. It must also be emphasized that, in these preliminary stages of development, it is simply impossible to conclude which of them are most promising and how great a contribution they can be expected to make to our future energy requirements, particularly in view of the fact that the Reagan administration appears to be deemphasizing much of the work on liquid fuels.

Enhanced Oil Recovery

We have seen that conventional methods of oil production recover only about one-third of the total amount of oil present in a reservoir. Obviously, the most effective way to increase our production of liquid fuels would be to improve the percentage of oil recovered from existing reserves. Various methods of enhanced oil recovery (also known as *improved* or *tertiary* recovery) are now under investigation and can be divided into three general classes.

1. Thermal Methods. These methods are used to decrease the viscosity of the oil by heating the oil-bearing strata of rock and thereby promoting better flow to the producing well. The most promising approach appears to be the direct injection of steam into the reservoir. This can be done in a cyclical way (often referred to as the *huff-and-puff* method) whereby steams is injected for a few days and then turned off while the accumulated oil is pumped out of the well. Another way is to inject steam continuously into an *injection well* so as to drive the oil through channels in the reservoir to a nearby *production well.*

Another thermal method is to inject air into the well to burn part of the oil, thereby furnishing heat that promotes the flow of oil to neighboring production wells. This method is hard to control and has not shown good results.

2. Miscible Methods. These methods involve the injection of a material that dissolves in the oil to form a less viscous mixture which can flow through the reservoir more easily than the original crude. Alcohols and light petroleum gases, such as propane and butane, have been tried, but the best appears to be gaseous carbon dioxide which is cheap and often available in large quantities in the oil fields. Miscible methods are fairly expensive, and it is necessary to recover and reuse as much as possible of the injected material.

3. Chemical Methods. These methods involve the injection of a material that reacts chemically with the constituents of the oil reservoir. For example, if a strong detergent solution is injected, it can emulsify the oil in the same way as in ordinary laundry operations. Another possibility is to modify water injection by adding a polymer to increase its effectiveness in displacing the oil. These methods are the newest and are quite expensive.

It is also possible to use a strong alkali, such as sodium hydroxide, with water flooding, since this reacts with the constituents present at the rock-oil interface and thereby promotes the release of the oil. Preliminary tests of this method are encouraging, but the method is obviously going to be very expensive.

Enhanced oil recovery accounts for less than 5 percent of our total production today. About two-thirds of the present *enhanced oil* comes from steam injection and about one-third from gas injection. It must be remembered that all methods of enhanced recovery are expensive. The extent of their use in the United States will depend strongly on the value of the recovered crude. Under optimal conditions, it has been estimated that enhanced recovery may contribute as much as one million barrels a day by the end of the 1980s—that is, some two and a half times today's production by this means.

Oil Shale

Oil shale is a carbonaceous rock containing an insoluble organic compound known as *kerogen.* Tremendous quantities of oil are potentially available from the oil shale which occurs widely in Colorado, Utah, and Wyoming. Although considered to be a new source, oil was actually being distilled from eastern shale prior to Colonel Drake's oil well in 1859. The western deposits were discovered in the early twentieth century, and possible production of oil from this shale has been considered from time to time for the past 75 years. A 1918 report by the United States Geological Survey discussed the growing production of oil in the United States and commented that it might eventually reach tremendous figures "because of the shale resource."

If oil shale is heated to about 500 degrees centigrade, the kerogen decomposes to give an oil similar to light crude oil. High-grade oil shale contains up to 30 gallons of oil per ton of rock. Estimates of the total amount of oil available from shale range from the hundreds to the thousands of billions of barrels, a quantity several times the total amount of oil available from our conventional oil fields. The overriding question is, "How do you get it out economically and without destroying the environment?"

Most of the commercially important deposits of oil shale are on federal land in relatively arid and unpopulated regions of the West. The shale occurs several hundred feet or more underground and must be mined by expensive underground methods and then brought to the surface for processing. A plant to produce some 50,000 barrels of oil per day (less than 1 percent of our present oil requirements) would necessitate an investment of close to $1 billion, would consume 5 to 10 million gallons a day of water, and would create some 60,000 tons a day of *spent shale* which has absolutely no value at all. New roads, sewers, schools, and towns would have to be built to accommodate the miners and construction workers. To fill 10 percent of our oil needs, the shale-mining industry would have to be as large as the entire coal-mining industry is today.

Very severe environmental restrictions would have to be met. When the shale is crushed and heated, a fine powder is released that may cause a dust problem. Millions of tons of spent shale would have to be disposed of in one way or another every year. The spent shale is alkaline and has a volume about 20 percent greater than that of the original shale (the expansion during heating is referred to as the *popcorn effect*). Phosphorus and nitrogen have to be added to spent shale if it is to be used for plant growth. The discharge water is contaminated with mineral salts which would impose a threat to all downstream water users, including the rich agricultural valleys in Arizona and California.

Because of the difficulties of aboveground processing, an underground *in situ* combustion method is also being investigated. This involves use of a slow-burning underground fire directly in the shale deposit to release the oil. This method uses less water and causes less environmental damage, but it is hard to control, and there are many questions about its commercial feasibility.

For many years, engineers have been saying, "Oil from shale will be economic if the price of crude goes up by 20 percent." Yet, even today, it is entirely possible that oil from shale may be some 20 percent more expensive than conventional oil. Despite its many problems, however, the potential oil shale resource is certainly too great to ignore. Federally funded demonstration plants are now in the design and construction stage. If these are successful, oil from shale may start making a modest contribution to our energy needs by the 1990s.

Tar Sands

Tremendous quantities of tar sands occur throughout the world near the margins of large sedimentary basins. These contain heavy unsaturated asphalt-type oils that have fairly high concentrations of sulfur, nitrogen, oxygen, and various trace metals. The most extensive deposits are in the northeastern section of Alberta, Canada, particularly the Athabasca deposits which have been known since 1788. Small deposits occur in Utah, California, and elsewhere in the United States, but to date these have not been considered important enough to develop commercially. The Alberta deposits alone are estimated to have some 280 billion barrels of recoverable oil.

Tar sands in Europe have been exploited for many years as a source of *bitumen* (asphalt) for road oils. Recovery is fairly simple: tar sands are simply mined by conventional surface mining and extracted with hot water to recover the bitumen, which can either be used directly as road oil or else distilled and then hydrogenated to produce a synthetic crude oil. Considerable sulfur is also recovered as a by-product.

The Alberta tar sands have been mined commercially since the late 1960s, when Suncor, Inc., and Syncrude Canada began production of 45,000 barrels per day of heavy crude oil which was converted by conventional refining operations into gasoline, fuel oil, and other petroleum products. The project was unprofitable until 1978, when rising world oil prices turned things around. At one time it was expected that Canadian tar sand production would reach some 500,000 barrels a day by the mind-1990s; these projects, however, are now encountering both political and financial problems, so that there is some question as to the magnitude of future production.

Liquid Fuels from Coal

In 1913, a German chemist, Friedrich Bergius, showed that coal could be converted to a synthetic liquid fuel by reacting it with hydrogen at high pressures. With very limited sources of petroleum, the Germans under Hitler operated twelve large *coal hydrogenation* plants which produced some 200 million barrels of liquid fuel a year during World War II. The process involved the direct reaction between coal and hydrogen at a temperature of about 800 degrees Fahrenheit and a pressure of several thousand pounds, using a catalyst such as cobalt or molybdenum; it produced a high-octane gasoline which was used for aviation fuel.

A second German process for the production of liquid fuels was based on the Fischer-Tropsch synthesis and became known as the *synthol* (synthetic oil) process. It involved the reaction between carbon monoxide and hydrogen in the presence of a cobalt catalyst; it could be modified to produce either a synthetic gasoline or a synthetic fuel oil. Six plants were operated in Germany during World War II with a combined capacity of four million barrels a year of liquid fuels.

Following the war, an attempt was made in the United States to commercialize the synthol process using cheap natural gas as raw material rather than coal. A plant was built at Brownsville, Texas, using an "improved" fluid catalyst. It was a scientific and commercial failure, never achieving more than a small fraction of its design capacity and proving completely uneconomic for the production of synthetic gasoline or fuel oil.

The one known successful application of the synthol process has been in South Africa, which is rich in minerals but has essentially no petroleum. As early as 1935, South Africa acquired rights to the German process and began active research on synthetic fuels. After World War II, a large plant was built at Sasol to produce synthetic gasoline and fuel oil from low-grade South African coal using an activated iron catalyst. Following successful operation of this initial plant, a much larger one (Sasol 2) has recently been brought into production. With Sasol 3 now under construction and completion expected by 1982 or 1983, most of South Africa's needs for liquid fuels will soon be met by synthetic fuels from coal.

The production of liquid fuels from coal is now being widely investigated in the United States as part of a $2 billion Department of Energy synfuels program. Most processes under development are modifications of German and South African technology. Very active research is under way by practically all the large United States energy companies, and the situation is changing almost on a day-to-day basis.

Most of the processes begin with the gasification of coal by means of the proven *Lurgi process,* which involves complete gasification of coal using a mixture of oxygen and steam under pressure. Gas leaving the furnace consists largely of carbon monoxide, carbon dioxide, and hydrogen. Solid materials are removed by filtration, sulfur dioxide is removed by conventional absorption, and the purified *synthesis gas* can then be converted to liquid fuels by three general processes.

1. Synthol Process. Based largely on the South African Sasol technology, the synthol process passes the purified synthesis gas over a fluid-bed catalyst to produce synthetic gasoline and fuel oil. An appreciable fraction of the synthesis gas is also converted to methane, which can be separated and used as a substitute natural gas. Other by-products include sulfur, ammonia, ethylene, and oxygenated compounds such as alcohols and ketones.

The Department of Energy's synthol program has emphasized the use of western coal, which is cheaper and easier to gasify than eastern coal. This process has one substantial advantage, namely that the technology is well worked out and understood; however, it is far from cheap. The Sasol 2 plant cost over $2 billion and uses 40,000 tons of coal daily. South African officials have been very close-mouthed about the economics of the process, merely claiming that gasoline costs less than $2 per gallon. In view of the fact that South African coal is fairly cheap, labor costs are much less than in the United States, and various subsidies and tax breaks are given to Sasol, the use of this process in the United States might easily result in gasoline costing several dollars a gallon.

2. Coal Liquefaction. This process is based largely on German technology. Synthesis gas from the Lurgi gasifier is reacted with steam in the presence of a catalyst to produce more hydrogen; carbon dioxide is removed, and the resulting gas, now largely hydrogen, is reacted with coal under high pressure to form liquid products.

In the 1950s, Union Carbide Chemicals Company operated a small pilot plant that liquefied about 300 tons of coal a day at Institute, West Virginia. Several coal hydrogenation pilot plants have also been funded in recent years by the Department of Energy, and considerable work is being done by private industry. For example, in October, 1979, Ashland Oil Company began operation of a pilot plant at Catlettsburg, Kentucky, which produces 1800 barrels per day of synthetic crude oil; based on this experience, they are now designing a larger plant which is expected to produce 50,000 barrels a day of crude.

A coal liquefaction plant designed by Exxon Research and Engineering Company is now in the start-up phase at Baytown, Texas. Employing the *donor solvent process,* powdered coal is mixed with a heavy oil and converted to synthetic liquid fuels by reacting it with hydrogen gas at high pressure. At full capacity, the plant will produce 600 barrels per day of synthetic oil.

In July, 1980, plans were announced for the world's first commercial-size plant to produce liquid fuels directly from coal. Using the *solvent-refined coal process* developed by Gulf Oil, the $1.4 billion plant will be located near Morgantown, West Virginia, and will produce about 20,000 barrels per day of a heavy oil that can be refined to gasoline and different grades of fuel oil. The project is being jointly financed by the United States Department of Energy and the governments of Japan and West Germany, who are interested in learning more about the technology of liquid fuels based on coal.

Coal liquefaction is more efficient in the utilization of hydrogen than the synthol process and produces fewer by-products. One major disadvantage is that western coal is not particularly suitable for liquefaction because of its higher content of oxygen and ash, so that most of the work is directed toward use of more expensive eastern coal. In addition, many of the chemicals present in coal liquids are extremely potent carcinogens such as the notorious 3,4-benzo(a) pyrene, which was found to be present in the air in several parts of the Union Carbide pilot plant. Despite rigorous protective measures, skin cancer reportedly occurred sixteen times as frequently among plant workers as it did among a comparable population outside the plant. Although good design can minimize such problems, environmental considerations will certainly add substantially to the cost of liquid fuels produced in this way.

3. Synthetic Methanol. Methanol, also known as *methyl alcohol* or *wood alcohol,* can be produced in small quantities by the destructive distillation of wood. It can also be produced synthetically by the high-pressure catalytic reaction between carbon monoxide and hydrogen. This is the process used in the United States to produce over 6 billion pounds a year of methanol at a price of under 10 cents a pound. Although, in theory, virtually any organic material can be used to produce methanol, almost our entire United States production is based on natural gas as a raw material.

Methanol can be used as an additive to augment gasoline supplies, but it does cause problems with plastic and rubber seals. In addition, concentrations above a few percent may result in corrosion of lead, zinc, or magnesium parts. Methanol has a lower heating value than gasoline, and the use of pure methanol requires a specially designed engine. Also, methanol based on natural gas does nothing for our energy resources, since natural gas itself suffers periodic shortages. For these reasons, methanol has not been widely discussed as a possible substitute for our liquid fuels.

Nevertheless, two developments in recent years have led to renewed interest in methanol. First, it can be readily produced from coal, using the conventional Lurgi gasifier to produce a hydrogen-carbon monoxide mixture, which, after purification, can be converted directly to methanol. Second, a new process developed by Mobil Oil Company chemically changes methanol to high-octane gasoline merely by passing it over a zeolite catalyst at a moderate temperature and pressure. This eliminates the problems associated with the use of methanol as a fuel and makes the methanol route to gasoline look more attractive.

It is also possible to use wood or agricultural wastes for the production of methanol, although a new type of gasifier would have to be developed. This would have the tremendous advantage of making possible a way to produce liquid fuels using domestic renewable resources. A 1979 report prepared for the Department of Energy estimated a maximum production of some 150 billion gallons a year of methanol could be achieved by the end of the century using just our renewable resources; however, it is very unlikely that substantial quantities of methanol will be produced from these materials until our coal reserves are depleted.

Ethanol and Gasohol

Another substitute liquid fuel, which has been widely investigated in recent years, is ethanol, also known as *ethyl alcohol* or *grain alcohol.* The mixture of 10 percent ethanol and 90 percent gasoline is known as *gasohol* and is already widely available in the United States. Ethanol has two substantial advantages over methanol: it has a higher heating value (giving more energy), and there are fewer corrosion problems than in the case of methanol. Addition of 10 percent ethanol to gasoline improves the octane number by two or three points, requires no modification of the engine, provides better performance than straight gasoline, and gives equivalent or even improved mileage.

One disadvantage of ethanol is its cost, which is two or three times that of methanol. A second disadvantage is that ordinary distillation produces a 190-proof ethanol which still contains 5 percent water. If this is used as a gasoline additive, the water may separate out and possibly freeze. The use of *anhydrous* (water-free) ethanol to produce gasohol is more expensive and requires more energy to manufacture.

Over 1 billion pounds of ethanol are produced in the United States yearly, most of it synthetically based on petroleum as a raw material. Obviously this method of producing ethanol does nothing to help our energy resources! Substantial quantities of natural ethanol are also produced by fermentation, but the cost is higher than the synthetic material; in the past, most fermentation ethanol went into beverage alcohols, but increasing quantities are now being used for gasohol. Much publicity has been given to the fact that any food or waste agricultural product containing carbohydrates can be used for ethanol production. For example, 1 bushel of corn produces about 2.5 gallons of ethanol, and a good deal of today's gasohol comes from ethanol produced from our "surplus" midwestern corn. It is also possible to use wheat, potatoes, sugar cane, and similar materials for ethanol production.

The production of ethanol from foodstuffs such as cereal grains is not justified. Considering the worldwide shortages of food that now exist and which will become increasingly common in future years, there is considerable question as to the ethics of using food to produce a fuel for automobiles. An even more impelling reason, however, is that the production of food in the United States is strongly energy-dependent. Gasoline and diesel fuel are used in the planting, cultivation, and harvesting of crops; fertilizers and insecticides are themselves made from petroleum or natural gas and require considerable energy in their manufacture; large quantities of energy are required in the recovery and purification of the fermented alcohol. This total amount of energy is greater than the amount of energy obtained when the ethanol itself is burned!

If ethanol is to make any real contribution to our future energy needs, two things will have to be done. First, we are going to have to use *waste agricultural streams,* particularly cellulosic material such as wood and cornstalks (the enzyme *cellulase* can be used to convert cellulose to glucose, which can then be fermented to ethanol in the usual way). Second, we are going to have to reduce the amount of energy used to produce the final ethanol. Considerable research is now under way to reduce both the amount of energy used in the distillation to produce 190-proof ethanol and the considerable energy required to remove the remaining 5 percent water and produce pure anhydrous alcohol. If this can be achieved, renewable domestic resources could serve as the source of at least part of our liquid fuels.

While large amounts of ethanol are being produced today in the United States from "surplus" corn and other grains, the most intensive work is being done in Brazil, where tremendous land areas are now planted with sugarcane for the production of ethanol. By burning waste sugarcane stalks (known as *bagasse*) to provide energy for purifying the ethanol, the process has been made highly energy efficient. With only very small quantities of domestic petroleum, Brazil hopes to produce enough ethanol to become largely independent of oil imports; its government officials plan to have at least one million of Brazil's seven million automobiles operating on pure anhydrous ethanol by 1982.

ALTERNATE SOURCES OF NATURAL GAS

While practically all the natural gas produced to date has come from porous sedimentary rock such as sandstone at depths of less than 15,000 feet, it is known that natural gas also occurs in certain geologic formations. These fall into five general categories: (1) western tight gas sands, (2) eastern Devonian shales, (3) coal seams, (4) geopressured brine, and (5) deposits of solid methane hydrates. Very large quantities of natural gas are present in these unconventional formations; however, there are doubts as to how much can be recovered using conventional technology—and what the cost would be.

Natural gas does occur in sandstone at depths greater than 15,000 feet, but the cost of drilling is much higher, and there is considerable uncertainty as to the availability because less is known about the geology of the deep reservoirs. To recover natural gas from low-porosity sedimentary rock such as shale or tight gas sands, it is first necessary to fracture the rock to provide channels for the gas to escape. Even when this is successful, production is much less than from conventional wells. Very small amounts are now being produced from shale, and some gas is recovered from coal seams before mining (chiefly as a safety measure); it is very doubtful that these sources will help much in the immediate future.

The Texas-Louisiana basin contains geopressured brine, another unconventional source of natural gas normally found below 15,000 feet. Here the natural gas occurs primarily as a solution in water under high pressure and at a temperature of about 180 degrees centigrade. The water also contains unfortunately large quantities of dissolved minerals, particularly common salt. Although fairly large quantities of natural gas may be present, the recovery cost is high—not only because of the great depth of drilling but also because of the need to handle and dispose of large amounts of corrosive saline water. A test well in Brazoria County, Texas, gave disappointing results at high cost. The Department of Energy plans to drill up to twenty wells in Texas and Louisiana to better determine the economics of geopressured gas, and several large oil companies have modest research programs looking into possible future development of this source of energy. There are also some efforts being directed toward possible use of geopressured brine for the generation of electricity, with recovered natural gas as a by-product.

The recovery of natural gas from methanehydrates is being investigated in the Soviet Union, where the materials occur in limited quantities in Siberia. No deposits have been found in the United States.

Substitute Natural Gas

It now appears that the most promising future source is *substitute natural gas* (SNG) produced from coal. Synthesis gas produced by the conventional Lurgi process is purified and then simply sent to a catalytic converter that converts the carbon monoxide and carbon dioxide to methane. The final SNG fuel contains over 85 percent methane and has a composition and heating value very close to that of natural gas.

Several pilot plants were operated in the United States during the 1960s to demonstrate the commercial feasibility of this process; these showed that SNG was not economically feasible at the then-existing price of natural gas. During the 1970s, when shortages and higher prices of natural gas came along, several oil companies and gas pipeline companies announced plans to build SNG plants, most of them based on less expensive coal from the western strip mines. In the meantime, inflation pushed up estimated costs of construction and operation, questions arose about meeting environmental restrictions, and the "gas bubble" temporarily relieved the pressure on our supplies of natural gas. In addition, Washington was reluctant to allow higher prices for SNG even though it would obviously cost more to produce than gas from conventional sources—and most of these building plans were abandoned.

In December 1979, the American Natural Resources Company, the fifteenth largest coal producer in the nation, announced plans for a $1.5 billion SNG plant to be constructed in North Dakota, using locally available lignite. The decision to build this plant was made possible by approval of higher rates for SNG negotiated with the Federal Energy Regulatory Commission. Called the Great Plains Coal Gasification Project, it was to be funded by a consortium of gas pipeline companies and gas distributors. Even though there were doubts about the financial returns, the consortium felt that ". . . the national interest requires construction of the project to proceed without delay." While this appeared to represent our first genuine step forward in the production of gas from coal, the entire project is now in litigation and its future is very much in doubt.

RENEWABLE SOURCES OF ENERGY

For over 200 years, this nation has been behaving like a young man with a $200-a-week job who suddenly inherits $1 million from his rich uncle. Overwhelmed by sudden wealth, he giddily begins to spend several thousand dollars a week with no thought for the future. Only when his capital is almost gone does he realize that he will soon have to give up this spendthrift way of life and again live on his $200 a week.

Our fossil fuels have represented "energy capital" that accumulated over the period of millions of years, and which we inherited when the United States was first settled. We have seen that roughly half our oil and natural gas are gone forever, with increasing inroads being made on our remaining coal reserves. At our present rate of consumption, our energy capital will be entirely gone in a few generations, and we will have to live on energy income. It is certainly time to take a closer look at renewable sources of energy and to find out how much they can be expected to help us get through the difficult years ahead.

Solar Energy

Much has been written during the twentieth century about solar energy and how it can contribute to our energy needs. The fact is that solar energy can make an increasingly important contribution to domestic heating and hot water requirements; however, it is very unlikely that solar energy will make a real contribution in the near future toward our two most important energy needs—electricity and liquid fuels. Solar energy suffers from several very serious disadvantages. It is low-level heat energy which requires enormously large collectors to yield any substantial total amount of energy. The energy available varies with the time of day, the time of year, weather conditions, and the latitude or distance a given location on earth lies above or below the equator. Tropical countries, which have no need for solar heating, receive the most intense solar radiation. Colder countries, such as Canada and those in northern Europe, receive very small amounts of solar heat for many months of the year.

The total amount of the sun's energy that falls on the earth can be calculated from the *solar constant* and the projected area of the earth. The solar constant (approximately 1.36 kilowatts per square meter) is simply the rate at which energy falls on a horizontal surface located at the average distance of the earth's atmosphere from the sun. This corresponds to some 172 trillion kilowatts continuously falling on the earth's atmosphere. (For purposes of comparison, the total installed generating capacity of all electrical utilities in the United States is about 525 million kilowatts, and the total amount of *all* energy used in the United States is equivalent to about 2500 million kilowatts.)

At first glance, then, it appears that solar energy might easily meet all the world's energy needs; but we should take a look at the fine print. On the average, about half of the incoming solar energy is lost by reflection back into space. Another substantial fraction of the energy that travels down through the atmosphere is absorbed by clouds, dust, smoke and haze, never reaching the surface of the earth. Since this surface is roughly three-quarters water, only one-quarter of the solar energy that reaches us falls on the land. This means that only about 6 percent of the earth's total solar energy is readily available for use.

Looked at from the standpoint of the United States, solar energy is even less promising. Averaged out over an entire year, the total amount of solar radiation reaching the land area of the United States is about half a trillion kilowatts, some 200 times our present total consumption of all forms of energy. Because solar radiation represents only very low-level heat, high losses result when this energy is converted to a useful form. It is also necessary to provide some type of large-scale storage of heat to supply nights and sunless days.

A major difficulty is the land area required for solar collectors. Modern electric generating plants have a capacity of about 1 million kilowatts; to produce this average amount of electricity, solar collectors would cover an area of some 20 square miles even if they could operate with the same efficiency as a modern power plant, which is quite unlikely. The electrical generation would vary from several million kilowatts during a sunny day in June to nothing at all at night and on cloudy days. A tremendously expensive installation would thus be used only a quarter of the time, and it would be fairly unreliable—with electric production coming to a halt every time the sun went behind a cloud. These and other problems in the possible use of solar energy for electric power generation are discussed later in this chapter.

Biomass

Another way to utilize solar energy is to convert it by photosynthesis into plant material (*biomass*), which can then be used as a source of fuel. Wood, of course, was the first fuel used by mankind and even today serves as an important fuel in many parts of the world. For example, it is estimated that wood supplies 11 percent of Vermont's energy needs at present, with a potential for an increase up to 25 percent.

An important consideration in our use of wood is the proper management of the thousands of woodlots spread across our lands. At present, most of these have either no management or else very poor management. By selective weeding, cutting, and harvesting, tremendous quantities of wood can be made available for fuel, while simultaneously promoting the growth of the remaining timber for sawlogs. It seems inevitable that sooner or later more controls are going to have to be placed on the management of our timber reserves. Wood will undoubtedly continue to supply a part of our energy needs, particularly as the fossil fuels become more expensive and less available; when these are eventually exhausted, we shall have to depend on wood for many of our energy needs.

The chief problem with biomass is the very low efficiency of photosynthesis. For example, sugarcane is one of the most efficient plants for utilizing solar energy, but only about 1 percent of the incoming energy is captured. Assuming that we could develop a crop that would grow throughout the United States and which would have an efficiency for photosynthesis equal to that of sugarcane, if the entire productive land area of the nation were planted with this crop (leaving little or none for food crops, highways, homes, or cities), the available energy would roughly equal our total present yearly consumption of energy!

Probably the most promising use of biomass in the United States today is to burn garbage or waste agricultural residues to produce steam. Several municipalities have constructed garbage incinerators that not only recover energy but also greatly reduce the volume of garbage and simplify the problem of disposal. Several of these have been quite successful. One question yet to be answered concerns the composition of the exhaust gases; our present garbage contains polymers such as vinyl chloride and polyurethane, and when these are burned they can produce toxic and/or carcinogenic materials. Problems have developed with several of the first of these incinerators to be operated. For example, Combustion Engineering Associates built a plant at Bridgeport, Connecticut, to convert municipal garbage to a black powdery fuel that could be burned to produce steam. The process used sulfur as an embrittling agent, and there were soon local protests about odors "bad enough to gag a maggot." Tests also indicated the presence of toxic dioxin (one of the chemicals associated with "Agent Orange") in the gaseous effluents, and the plant was closed for modifications.

It is also possible to convert animal and agricultural wastes directly to a methane-containing gas that can be used for fuel. This is already being done in some municipal sewage-disposal plants, where the raw sewage is blown with air that oxidizes some of the material to form methane. Manure from feedlots can also be used to produce a *biogas* by simple decomposition in the absence of air, and at least one western plant is now doing this commercially.

Solar Heating

For at least the next twenty years, the most promising use of solar energy appears to be for domestic heating and/or hot water. Two general types of systems are being widely investigated. In *passive* solar heating, there are no moving parts; the heating results merely from the location and landscaping of a particular structure, which becomes in itself a large solar collector. For example, a building can be faced to the south so that the sun's heat is captured through large double- or triple-glazed windows. Means for storage of heat may be provided by using Trombé walls or black floor tiles to soak up heat, which is slowly released after the sun goes down. In some cases, an overhanging roof may be used to cut off the high (and hot) summer sun and yet permit entry of the lower winter sunshine.

Passive solar heating has been shown to be both effective and inexpensive. Substantial savings in the cost of heating can be realized by combining passive solar heating with other means of heat conservation, such as better insulation, storm windows, weather stripping, and the like. Passive solar heating suffers from one major drawback—it is practical only for new structures; thus the effect of passive solar heating will be felt only as new buildings replace the old, and then only if it is included in most of the new ones.

In *active* solar heating, large south facing panels of glass, plastic, copper, or aluminum are placed on the roof of a building to capture the sun's energy. Water from a large insulated storage tank may then be circulated to the collectors and either used directly for hot water or else used to heat the building by pumping through conventional radiators. A second option is to circulate air to the collectors, which can then be used for heating in the same manner as a hot-air furnace. If storage of heat is desired, it can be provided by circulating the air through a large rock pile to soak up heat. Active solar heating can be used in any building, new or old. Its chief disadvantage is the installation cost, typically $2000 or $3000 for even the most inexpensive system. If a homeowner has to pay 15 percent interest on a home-improvement loan, the expected savings from solar heating may not even cover the interest.

There are also other problems with all solar heating systems. A backup system is required for cloudy days, which means there will be no saving on the installation of a conventional heating system although, of course there will be saving in the cost of fuel. Local building codes may place a barrier in the way of solar heating, and property taxes may even be increased if solar heating is installed. Electrical utilities may refuse to permit lower rates for backup systems in buildings with solar heating. In addition, since solar heating is fairly new, many contractors who install the systems are inexperienced and, as a result, there have been many unexpected problems and high maintenance costs associated with units installed in recent years.

It is obvious that future use of solar heating in our homes is going to depend very largely on whether or not there are sufficient economic incentives to justify its use. Passive solar heating should certainly be considered very seriously by anyone planning new construction. Many municipalities have already relaxed their building codes and eliminated taxes on solar installations. Federal income-tax credits are already available for certain types of solar collectors, and these should be continued and possibly expanded. Eventually some regulation of the industry is going to be required to reduce costs and improve the reliability of solar collectors. Optimists believe that as much as 40 percent of our energy could come from solar by the end of this century—provided we make the proper moves now. The editor of *World Oil* says that solar energy will have the impact of "a mosquito bite on an elephant's fanny." The truth is undoubtedly somewhere between these two extremes.

ELECTRIC POWER: THE CHALLENGE

If the electric utility industry is to survive as we have known it in the United States, several basic changes will have to be made. In the first place, our state governors are going to have to realize that seats on the public utility commission can no longer be used as a way to pay off political debts. The regulatory process is now quite complex and demands a knowledge of economics, political science, sociology, environmental aspects, as well as some technical understanding of the engineering principles involved in the generation and distribution of electricity. Our entire way of life is based on electricity, and the electrical utility industry is now in trouble; decisions concerning the future of electric power—both in the state capitals and in Washington—are too important to leave to any but the most highly qualified individuals that can be found to serve.

It must also be recognized that the privately owned utilities are operated for one chief purpose: to make as much money as possible for their stockholders. Since their return is, at least in theory, based primarily on their investment, their philosophy has been consistently growth-oriented. As a result, use of electricity has always been encouraged in any way possible, regardless of waste or need. Newspaper and magazine advertisements have shown us over the years the wonderful gadgets and labor-saving devices that could be run on electricity; it was implied that a homemaker who did not have a majority of these aids was truly "deprived." Electric dryers, ranges, heat, color televisions, and air conditioning were widely promoted. The symbol of American affluence might well be the self-cleaning electric oven operating at 700 degrees Fahrenheit in an air-conditioned house!

How effective has the federal government been in reducing the use of oil and natural gas for the generation of electricity? The use of natural gas has remained essentially constant since the Arab embargo. The use of petroleum actually increased—and in 1978 was some 20 percent higher than in the 1973–74 period! There was a drop in consumption in 1979, but this was partly the result of a milder winter. It is obvious that government pressure to reduce the use of these two critical fuels has not been very successful.

As we mentioned earlier, after the Arab embargo of 1973–74, there was considerable uncertainty about the future growth of the industry, and the utilities canceled or significantly delayed construction of much new capacity, mostly nuclear or coal-fired, then on the drawing boards. For a few years, it was quietly assumed that, if the increased demand did materialize, it would be met by installing new gas turbines which would operate on residential-type heating oil. What was the reasoning behind this? Gas turbines can be installed in roughly two years, at less than half the cost per kilowatt of capacity of a nuclear or coal-fired plant. There is no hassle over environmental effects, radioactivity, or nuclear wastes. Even though gas turbines use expensive and critical petroleum or natural gas, the higher fuel costs could immediately be passed along to the consumer through the fuel-adjustment clause. The fact was that the "bottom line" of utility company financial statements dictated the use of gas turbines even though it would have required higher oil imports and increased vulnerability to possible future oil embargoes. Such plans as this were, of course, shelved by President Carter's energy program.

It is absolutely essential that the prime goal of the electric utilities be redirected toward minimizing the use of electricity, particularly the use of oil and natural gas. As the industry discontinues its hard sell of electricity and begins to place first importance on conservation, some of the aspects that regulatory agencies and commissions together with industry leaders should seriously consider are the following.

Electric Heat

The utilities have been very successful in selling electric heat, and in some areas such as New England the maximum system demand occurs during the winter heating season. We have seen that only about one-third of the heat energy of a fuel can be turned into electricity; if the same fuel is burned in a coal- or oil-fired furnace, roughly two-thirds of its heat energy can be utilized for space heating. Not only is electric heat a wasteful use of electricity; it contributes significantly to power peaking which, in turn, requires use of fuel oil or natural gas in backup generating facilities. The use of electric heat in any new construction must be banned unless there is good reason other fuels cannot be used.

Voltage Reduction

The rate at which power is delivered to a motor or an electrical appliance depends on the applied voltage. It has been found that a utility can reduce the demand for electricity by lowering voltage as much as 5 percent. Lights are slightly dimmed and motors will run a bit longer, but no damage results—and the average person does not even notice a change.

Voltage reduction has been used occasionally under emergency conditions to handle peak loads, usually during the summer months when generating facilities were inadequate to handle the high air-conditioning load. The success of voltage reduction depends to a large extent on the nature of the load on the system, and it may not be advantageous in all cases. It does, however, offer a way to reduce or at least "iron out" peak demands and thereby conserve the oil or natural gas otherwise used for peaking power. All electric utility systems should be analyzed to determine the effectiveness of voltage reduction; where it can conserve energy, voltage reduction should be a routine procedure during all periods of high demand—not just as an emergency measure.

Air Conditioning

In much of the United States, the maximum demand for electricity occurs during the summer air-conditioning season. Growing use of air conditioning in private homes, in large windowless office buildings, and in large indoor shopping malls is consuming tremendous amounts of electricity. Often the same area that is heated to near 80 degrees Fahrenheit in the winter is cooled below 70 degrees Fahrenheit in the summer. When consumers begin to realize that the comfort from summer air conditioning may require them to pay billions of dollars for additional generating facilities, perhaps they will become more tolerant about putting up with a few uncomfortable days in the summer and enjoy fresh air again.

Manufacturers are now required to specify the energy efficiency of electrical appliances such as refrigerators and air conditioners. There is no similar requirement when it comes to new building construction. An environmental impact study is often required to obtain a building permit; a similar *energy impact study* should also be required. This should specify the energy demand of each new facility, including those steps taken to minimize energy consumption. *No building permit should be issued unless it can be shown that the building has been designed in accordance with approved methods of energy conservation.*

The Electric Rate Structure

For years, electric utilities have used a pyramid rate structure whereby large consumers of electricity pay less per kilowatt-hour than small consumers. While this was undoubtedly done to encourage the use of electricity, there was also some justification for these rates: since the large-scale generation and transmission of electricity is more cost-efficient and since some of the fixed costs are independent of the total quantity of electricity used, it is true that total costs do increase more slowly than total consumption.

In recent years, these *quantity discounts* have some under increasing pressure from both consumers and environmentalists who argue that they no longer reflect the true cost of electricity. Environmentalists argue that the economic and social costs of producing electricity should be included in the rates, and that quantity discounts only encourage environmentally wasteful consumption by large users. Consumers argue that inflation has made new generating facilities frightfully expensive in comparison to existing facilities, and hence the additional electricity should bear a higher price tag. (This argument is not necessarily valid, because transmission and metering costs are only loosely connected to the total consumption; the real cost of the additional electricity could actually be lower.)

Another factor that should be taken into consideration in the pricing of electricity is the problem of power-peaking. It has been estimated that the actual cost of a kilowatt-hour of electricity supplied to meet peak demand may be as much as *twenty* times the cost of a kilowatt-hour during nonpeak demand. Nevertheless, utility commissions have, with few exceptions, established electric rates simply by dividing the electric utility's total yearly costs (including a reasonable return to its stockholders) by the total yearly output in kilowatt-hours. This effectively averages the cost of electricity, so that the nonpeak user is paying too much and the peak user is paying too little. Even worse, it has encouraged the rapid growth of peak-power demand.

In recent years, there have been sporadic attempts to encourage off-peak use of electricity by offering lower night rates combined with higher day rates. These have found some success in industry, where the electric bill is often a large part of the total cost of doing business. Off-peak rates, when available, have been pretty much a failure when it comes to residential use of electricity. They require a change in individual habits such as doing the laundry and ironing just before bedtime and using the electric water heater only in the evening. Possible savings are quite small after the cost of the special metering and bookkeeping required is passed along to the consumer, and the average homeowner is not about to change his or her life-style just to save a dollar or two on the monthly electric bill.

What we must remember is that the energy crisis is here to stay as long as most of our electric power comes from the burning of fossil fuels. With this in mind, the most important consideration in establishing a rate structure is that it should encourage and reward conservation. Unfortunately, this is directly contrary to the philosophy of the electric utility industry since it began some hundred years ago. Without the continual addition of new generating facilities, there will be no increase in the "return on investment," and stockholders can look forward only to static or even declining earnings as higher fuel costs put greater pressure on profits. The utilities are literally being asked to kill the goose that lays the golden eggs.

Since the pyramid rate structure encourages wasteful use of electricity, it should be abolished and replaced by a flat rate structure. In fact, some consumer groups recommend a "reverse pyramid," whereby large consumers would pay increasingly higher rates per kilowatt-hour as consumption goes up. A promising approach has been taken by certain utilities, such as Pacific Gas and Electric, which have established "lifeline" rates to encourage energy conservation by residential consumers. These permit the use of some 240 kilowatt-hours per month at a low subsidized rate, with all excess being charged the usual rate which is some 50 percent above the "lifeline."

The second thing we can ask of a rate structure is that the price of electricity represent at least a close approximation to the cost of producing that electricity. In particular, something must be done about the great discrepancy between peak and nonpeak costs. Substantially higher rates may have to be charged during peak hours, and industry may have to shift some of its load to evening or weekend hours.

One sure way to reduce consumption of electricity is to substantially increase the rates. This has proven to be very effective in reducing the consumption of gasoline and fuel oil during the past two years when the federal government more or less encouraged the oil companies to raise prices as a means of conservation. During this same time, of course, the utilities have faced only increasing opposition to rate increases. Since an electric utility company is subject to public regulation, it is far more vulnerable to consumer pressure than a large oil company such as Exxon. It makes no sense to talk of higher taxes on oil as a means of conservation without at the same time considering a similar tax to encourage conservation of electricity. Perhaps all utilities should be required to establish "lifeline" rates—with a substantial federal tax on every kilowatt-hour above the "lifeline."

Facing the United States is the serious question of whether its many state utility commissions will be able to make decisions that encourage conservation and the wise use of electricity. Consumers and environmentalists want low electric rates but without additional environmental degradation. The utilities want more money for their stockholders. It is hard to believe that these conflicting interests can best be resolved by increased government regulation.

On the other hand, many Americans now feel that the electric utilities should be taken over by states and municipalities, since these can obtain funding by issuing tax-exempt bonds which are more easily marketed and bear lower interest rates; this would, of course, be an indirect subsidy. Others have gone even farther by suggesting that the federal government purchase utility securities to provide funding for their substantial construction costs. There has even been talk of nationalizing the utilities, as has been done in practically every other nation in the world. The entire electric utility industry faces a very real challenge today; its problems can and must be resolved in a way that is consistent with the interests of the consumer, the environmentalist, and the future security of the United States.

OTHER POSSIBLE SOURCES OF ELECTRIC POWER

In 1979, coal, hydro, and nuclear accounted for almost three-fourths of the total electric generation in the United States. To some degree, hydro is a renewable resource and presumably will always be available to us. There is enough coal to last for several hundred years at the present rate of consumption and enough uranium to last well into the twenty-first century, even without development of the fast-breeder reactor. There should be no major problem in supplying electricity over the immediate future if we can learn to practice conservation.

Several alternate ways to produce electricity are now being studied to determine whether or not they can contribute to our future needs. These involve strictly long-range research, and it is very unlikely that any of them will make a substantial contribution in the immediate future. Power plants of the twenty-first century will continue to be predominantly coal, hydro, and nuclear, but it is important to our perspective on the entire energy story that we have a realistic view of these alternatives.

Geothermal

If a hole is drilled into the earth at any point, it is found that the temperature increases by about 50 degrees Centigrade for each mile of depth. This geothermal energy comes from two sources: (1) The decay of radioactive elements existing in the earth's crust, and (2) the flow of heat outward from the molten magma that constitutes the inner core of the earth. The amount of energy potentially available is extremely large, but technical considerations make it very difficult to utilize thermal energy. For example, in 1979, geothermal energy accounted for less than 0.2 percent of the total production of electricity in the United States.

Several different types of geothermal energy are available. In some parts of the world, such as Iceland and the western United States, underground reservoirs of geothermally heated water are fairly close to the surface. If these are located close enough to a populated area, they can be tapped for hot water for space heating. Also some electricity is generated in Europe from such deposits; they have not been used for this purpose in the United States because their high content of dissolved salts (several times that of sea water) causes them to be very corrosive to generating equipment. In addition, there is a problem of disposing of the tremendously large amounts of saline water which must be brought to the surface. Since these hot-water reservoirs do represent a large fraction of the total geothermal resource, the Department of Energy is looking into possible economic ways of using them.

Very occasionally, the pressure in an underground reservoir may be so low that the water has boiled to form steam. This is the easiest and most economical form of geothermal energy to use, because corrosion is minimal and there is no need to handle the saline water. Pacific Gas and Electric has a geothermal power plant at The Geysers, near San Francisco, which produces about 500,000 kilowatts of electricity from geothermal steam. Because of the rarity of this type of reservoir, however, it is extremely limited as a source of electricity.

Most of the potentially available geothermal energy is present either in dry rock or in the molten magma core of the earth and is known as *sensible heat.* To exploit this source, it would be necessary to fracture the rock several miles underground to form a permeable network of channels, inject water into these channels and withdraw the steam at the earth's surface. Modest worldwide research is being carried out, but the future of this energy source is very doubtful because of the severe technological problems involved.

Although geothermal energy is not, strictly speaking, a renewable source, the potential is so large that it is practically infinite. It remains a tantalizing prospect but will not become an important source of power within the foreseeable future.

Solar Power

We have already considered the potential prospects for solar energy and some of the problems involved in its utilization. When we look into the possibility of using solar energy to produce electricity, it turns out that there are many different forms of solar energy that might be useful to us, including hydroelectricity (the only one now being used), solar cells, solar thermal conversion, wind power, ocean thermal energy conversion, space solar systems, and energy from tides, waves, and currents. In evaluating these, however, we must distinguish between those that are *interruptable* (that is, available only when the sun is shining) and those that would theoretically be available around the clock. Interruptable sources could make a contribution during such times as they were available, but it would still be necessary to have sufficient other generating capacity to meet the peak loads.

From the standpoint of government financing, the most active project is the *solar tower* concept, in which large numbers of adjustable mirrors (known as *helistats*) focus sunlight on a steam boiler located at the top of a large tower. A 10,000-kilowatt demonstration plant is now being constructed at Barstow, California, to test the general concept. To obtain any reasonable capacity, a fairly large land area is needed for the mirrors; in addition, substantial quantities of steel and concrete are required for the tower and solar-collecting facilities. The system would obviously have to operate as a supplemental power source, or else some type of pumped storage would have to be included to give greater availability. Very preliminary estimates indicate the cost of this electricity would be up to ten times that of conventional systems, and it appears very unlikely that the solar tower will make any real contribution to our needs for some time to come.

The solar process which now appears to be most promising is photovoltaic conversion of sunlight through the use of the so-called *solar cell.* This has been widely used in the space program, where solar cells have been used to power instruments located on the spacecraft. A solar cell consists essentially of small wafer-like crystals of a semiconductor such as silicon; when exposed to sunlight, these act like small batteries and produce a voltage difference. If the individual cells are hooked up in series to obtain the proper overall voltage, they may then be used to produce directly an electric current. There is no need for a steam boiler, turbine, condenser, pipes, pumps, and all the other gadgets used in the conventional power plant.

Solar cells do have several disadvantages. They are extremely expensive, for one thing, with the cost of electricity being over twenty times that produced by our conventional sources. The efficiency for producing electricity is fairly low, some 10 or 15 percent, but this may be increased by technical improvements in the manufacture of semiconductors. Large collecting areas are required to produce any substantial quantity of electricity, and some type of pumped storage would be needed for round-the-clock operation. Solar cells produce a direct current, which would have to be changed to alternating current before it could be distributed to customers. Despite these problems, however, solar cells offer many potential benefits which might be achieved through long-term research. It is possible that solar cells will make a modest contribution to our energy needs by the beginning of the next century.

Another idea (which might well be called "way out") for recovering solar energy is to use a space shuttle to place solar cells in outer space where they would be continuously exposed to intense sunlight, undiluted by passage through the earth's atmosphere. Electrical energy produced in space would be converted to a microwave beam that would be transmitted to earth and changed back to electricity. The whole concept is frightfully expensive, and there are potentially serious environmental effects associated with pollution of the upper atmosphere—and possible harmful biological effects from the microwave radiation that is beamed back to earth. Perhaps this idea is best left to the science fiction writers.

Ocean thermal energy conversion is another solar process and involves utilization of the different temperatures between the surface and deep waters of tropical oceans (approximately 40 degrees Fahrenheit) to drive a heat engine. Although tremendous quantities of energy are theoretically available, there are many practical problems. Because of the relatively small temperature difference, the thermodynamic efficiency is only some 3 percent or less. Large amounts of heat exchange fluid must be circulated, and large heat transfer surfaces are required for any appreciable power production. Certain marine organisms like to grow on the heat transfer surfaces, and this may cause problems of fouling. The one substantial advantage of the system is that it could be used for base-load generation of electricity without the use of storage facilities. Although a small demonstration plant was operated recently in Hawaii, and a larger unit may be tested some time in the 1980s, ocean thermal energy conversion will not contribute to our needs in the foreseeable future.

Power from *wind, waves, tides,* and *ocean currents* represents less direct forms of solar energy. Windmills have been used for centuries to pump water and to turn grindstones, and in recent years more attention is being paid to this source of energy. The total amount of energy associated with the winds of the world is very large indeed, and even if only 1 percent of it could be harnessed, it would supply several times the world's need for electricity. Despite the tremendous potential, actual use of wind energy has been very small. The greatest use was in Denmark during World War II when wind turbines produced an average power of about 300 kilowatts during a period when conventional fuels were in very short supply.

Various types of propeller- and turbine-driven machines have been designed to recover wind energy. They are fairly simple mechanical devices and can be connected to an electric grid to supply power. Several different types are commercially available, but the cost is quite high and they cannot compete at present with conventional power sources. The chief problem in the use of wind energy is that it requires an average wind speed of at least 12 miles per hour for any practical production of electricity. This severely limits the number of suitable sites. Since wind energy is quite variable, it can serve only as a supplemental source of electricity, or else pumped storage capacity would have to be provided. Since the individual units are fairly small, they must be spread over a considerable land area to prevent interference with each other; it has been estimated that a land area of several hundred square miles would be required to produce power equivalent to that of a large conventional utility plant. The rapidly spinning blades could kill large numbers of birds, and the noise could disrupt wildlife. The large blades would also interfere with television reception, and power lines would have to be run all over the landscape to connect individual units. It is hard to be optimistic about the future of wind power except for small-scale use in isolated situations.

Much smaller quantities of energy are associated with the tides, waves, and ocean currents. The tidal power plant presently located at the mouth of the Rance River in Normandy, France, has a capacity that varies between 0 and 240 megawatts (240,000 kilowatts), depending on whether the tide is high or low; average output is only about 70 megawatts. France's plans for the future include additional dependence on tidal power. In the United States, modest amounts of tidal power can be developed, but the potential is too small to have a significant impact on our future needs. Canada's Bay of Fundy appears to be the best source of tidal power in North America at present.

We have already discussed the possible use of *biomass,* the capturing of solar energy from some type of plant matter through the process of photosynthesis. The material can then be burned to produce steam, which can be converted to electricity using the conventional steam turbine. Several possibilities exist, such as the harvesting of trees, the raising of fast-growing plants on special "energy farms," or the utilization of organic materials in municipal and agricultural wastes.

A very small amount of electricity, less than 0.1 percent of the total United States production, now comes from the burning of wood and garbage wastes. The problem is essentially one of availability; even a rural state such as Vermont, which does produce some electricity from the burning of wood chips, would be very hard pressed to generate all its electric power in this way. There have been suggestions that energy farms be established on marginal lands, using fast-growing plants such as sunflowers, rubber plants, or eucalyptus trees. This would increase the availability of biomass, but it would present new problems. It is wasteful to simply burn materials that could be used either for food or for the manufacture of liquid fuels. Energy farms would compete with conventional agriculture for land, water, and fertilizers. Large plantations of the same crop would be very susceptible to disease, and there is a danger that any rapidly growing plant might spread too fast and wipe out the native plant life of an area.

Agricultural wastes include such things as the inedible parts of food crops, manures, and the unused parts of trees which are harvested for lumber or paper. The total amount of available energy is fairly small and many of these materials are widely distributed through rural areas. Also, much of the material is recycled to improve the fertility of the soil. If this were burned, it would be necessary to replace it with a chemical fertilizer which requires considerable energy to manufacture. Agricultural wastes should stay on the farm.

The use of municipal wastes (garbage) to produce steam or electricity is one prospect that holds real promise for the future. These wastes are readily available, since they are already being collected and transported to central locations. In addition, most municipalities are having difficulty disposing of these wastes, and they would probably be glad to pay someone to take them away. There is also the possibility of sorting these materials, either before or after combustion, to recover valuable noncombustibles such as steel and aluminum. A method of burning that will eliminate the problem of toxic materials will have to be found. While the amount of energy available from this source is fairly small, it would make some contribution and at the same time greatly minimize our solid refuse problem.

Fuel Cells

The concept of the fuel cell is far from new, but considerable research is now being directed toward its possible use to produce electricity. A fuel cell operates on the same principle as a battery—a chemical reaction that produces electrons takes place at one electrode and a chemical reaction that consumes electrons takes place at the other electrode. This builds up a voltage difference between the two electrodes which can be used as a source of power. In the fuel cell, the chemical reactants are supplied continuously from the outside and the reaction products are continuously removed; in contrast to a battery, a fuel cell can, in theory, operate indefinitely. The best chemical fuel is hydrogen, since this produces only water as a reaction product. There are also plans to eventually use a cheap hydrogen-rich gas made from coal.

Fuel cells have several advantages. They produce electricity without the need for boilers, turbines, condensers, and all their necessary adjuncts. They have a high thermodynamic efficiency, theoretically up to 60 percent, and this is maintained over a wide range of loads. They emit few pollutants and can be operated conveniently in large or small sizes. A 4500-kilowatt demonstration plant is now under construction in New York City adjacent to the Consolidated Edison Company's East River generating station; its thermodynamic efficiency is expected to be 37 percent, and it is hoped this can be increased to 40 to 55 percent with improved technology.

In practice thus far, fuel cells have not lived up to their expectations. They are costly, require substantial amounts of critical materials, have had short operating lifetimes, and their efficiencies are far lower in practice than in theory. If and when these technical problems can be overcome, it is possible that fuel cells may make an important contribution to our power needs by the twenty-first century.

Some enthusiasts believe that fuel cells may play an important future role in what they foresee as a *hydrogen economy.* They envision large-scale production and pipeline distribution of hydrogen; this would fuel our automobiles, heat our houses, produce our electricity via fuel cells, and serve as a raw material for the chemical industry—but it certainly is not something the consumer has to worry about in the near future.

Controlled Thermonuclear Fusion

The energy of the stars (including our sun) comes from *thermonuclear fusion* reactions, whereby hydrogen gas is converted to helium with the release of large quantities of energy. For some thirty years, considerable research has been directed toward the possible use of thermonuclear fusion reactions on earth for the generation of electric power. Many different approaches have been explored, and some progress has been made, but it has turned out to be far more difficult than was originally anticipated; there is still much doubt as to whether thermonuclear fusion will ever be a practical source of energy on earth.

Most of the fusion work has been directed toward the possible use of gaseous deuterium (heavy hydrogen) as a thermonuclear fuel. When heated to temperatures of some 100 million degrees centigrade or more, deuterium reacts to form helium plus other reaction products—with the evolution of considerable energy. To produce useful power, the gas must be confined by the use of magnetic and electric fields for a time long enough so that the energy produced is substantially more than that required to get the reaction going. To date, no thermonuclear fusion machine has even produced as much energy as was put in.

There are two major advantages of thermonuclear fusion: (1) Even the small amount of deuterium naturally present in water would provide almost unlimited fuel reserves. No longer would there be "have" and "have not" nations. (2) Thermonuclear fusion produces only small quantities of radioactive products, virtually eliminating the problem of radioactive waste disposal.

A substantial amount of money has been invested in fusion research; for example, Princeton's giant Tokamak Fusion Test Reactor is being constructed at an estimated cost of $240 million. Fusion reactors require powerful magnets and consume large amounts of electricity. Some proposals involve very sophisticated technology such as the possibility of cooling magnet coils to very low temperatures with liquid hydrogen. The engineering problems associated with the recovery and utilization of heat from a fusion reactor have not even been considered. Thermonuclear fusion remains a vague hope for the distant future.

CONSERVATION: OUR BEST "RESOURCE"

If we look at our use of energy in the United States today, we find that petroleum supplies 47 percent and natural gas supplies 26 percent of the total demand. Our most abundant fossil fuel, coal, contributes only 19 percent of our energy. Hydro and nuclear power each supply about 4 percent of our total demand, and all other minor sources together (such as wood, waste, and geothermal) contribute only about 0.1 percent. Note that petroleum is supplying almost half of our energy, with over one-third of this petroleum now supplied by imports.

A good picture of the wasteful American way of life can be had by looking at our total consumption of energy. Each year, we use energy equivalent to over 12 tons of coal per capita. West Germany, Switzerland, Sweden, Norway, and the Netherlands have standards of living just as high as ours but use only about 6 tons per capita—roughly half of our consumption. Japan, despite its remarkable postwar industrial growth, uses only one-third as much energy per capita as the United States. India uses one-fourth of a ton per capita, and Bangladesh, one of the really poor regions of the world, uses roughly 70 pounds per capita for all its energy needs. Many families in the United States use that amount of energy just for a backyard barbecue!

If we could reduce our per capita energy consumption to that of Western Europe, there would be no need for imported oil, no need for nuclear power plants, and our environmental problems would be greatly reduced. Some people insist that any attempt to reduce energy consumption in the United States must lead to a reduction in our standard of living; the fact that western Europe today uses half the energy per capita to achieve a standard of living equal to ours proves the fallacy of this idea. We can and we must reduce our consumption of energy, particularly our imports of crude oil. Recent events in Iran demonstrate only too vividly the danger of relying on unstable regions of the world for supplies of critical raw materials. When the Arabs carried out their boycott in 1973, we could quite easily have reduced our crude oil consumption by 5 or 10 percent and thus been able to tell them, in polite diplomatic language, exactly where to go. In the 1980s, it will be far more difficult to free ourselves of wasteful energy habits, but we must go full speed ahead regardless of political considerations and the difficult rethinking of our jet-lag attitudes.

The United States has about 5 percent of the world's population and also about 5 percent of the world's proved reserves of petroleum; yet we were using some two-thirds of the entire world production of petroleum through the early 1950s and even today are using about 30 percent. Our demand for petroleum products has more than tripled since the end of World War II. Not surprisingly, serious questions have been raised as to how much longer we can expect to use up such a high fraction of the world's resources to support our wasteful life-style. It is obvious that conservation must be our first order of business in the 1980s.

Chapter Five

Polymers and Minerals

For many years, the United States was using half or more of the world's annual production of fuels and minerals. Although our share had been reduced to about 42 percent by 1940, even as recently as 1952 we were using over half the world production of crude oil. With the worldwide industrialization that has occurred since World War II, our share has continued to decrease; yet, even today, the 5 percent of the world population who live in the United States are using some 27 percent of the total world production of fuels and minerals, including 30 percent of the petroleum, 30 percent of the aluminum, 41 percent of the lead, and 44 percent of the molybdenum. Since 1954, our country alone has consumed more minerals than the rest of mankind used in all previous history.

With the United States facing a genuine shortage of energy, a logical question is, "Are we also facing a shortage of essential materials?" The answer seems to be that the "material crisis" will be much less urgent than the energy crisis although it will be closely related.

Even though it is unlikely that we will actually run out of our basic raw materials in the foreseeable future, one thing is certain: they will become increasingly expensive. Large amounts of energy are required for the mining, processing, and refining of these materials. In addition, as it becomes necessary to utilize lower-grade ores, even greater amounts of energy will be required. Since it is virtually certain that the cost of energy will continue to rise, this will effectively increase the cost of just about everything else we use.

At the same time, other factors will act to control costs. Increased recycling of materials such as lead and aluminum will save a great deal of energy. Substitutes will undoubtedly be developed for many of our more critical and costly materials. Technological improvements will be introduced. As an example, practically all ores are now put through a flotation process whereby they are pulverized and mixed with chemical soaps or flotation agents to separate the desired mineral from the waste material or gangue; this permits use of lower-grade ores, produces a more concentrated product, and substantially reduces the cost of further refining operations.

In recent years, one of the most obvious aspects of our throw-away economy has been the increasing use of synthetic organic chemicals, particularly synthetic polymers, in our daily life. Most noticeable has been the explosion of packaging materials, which are now literally producing mountains of municipal garbage. Our overall use of polymers is now so widespread and has shown such rapid growth that it is hard to realize the extent to which we have become dependent on these materials.

The catch, of course, is that practically all polymers and other synthetic organic chemicals are being produced from petroleum and natural gas. The cost and continued availability of these materials, which have become so

Figure 21. New Yorkers maneuver through piles of garbage (mostly packaging materials) during a strike by sanitation workers. (Wide World Photos, Inc.)

important to us, depend on the continued availability of those same critical resources that are associated with our energy problems. It is very important to realize that our energy crisis includes a crisis in this supply of vital synthetic materials. This is far more serious than any future difficulty we may have in obtaining the ores and minerals needed to produce inorganic materials such as brick, concrete, glass, and metals.

In previous chapters we have covered the availability of renewable agricultural resources of animal and vegetable origin and the availability of renewable and nonrenewable fuel resources. This chapter considers two other types of nonrenewable resources: (1) Synthetic polymers, which are produced largely from petroleum and natural gas, and (2) the naturally occurring inorganic materials, such as sand, lime, salt, and the metal ores. The last half of this chapter concentrates particularly on those raw materials essential for the production of our metals.

PETROCHEMICALS AND POLYMERS

To understand the current synthetic organic chemical industry and the problems it is facing, we must go back several centuries to the days when chemistry was first emerging as an exact science. Two separate branches of chemistry gradually evolved: *inorganic chemistry,* which dealt with minerals and other chemicals present naturally in the earth, and *organic chemistry,* which dealt with those chemicals found in living plants and animals. For many centuries it was believed that the organic chemicals could be produced only through the action of a living organism. In 1828, however, a German chemist named Friedrich Wöhler accidentally synthesized *urea,* an organic chemical present in urine, from purely inorganic raw materials. This classic experiment showed that there was, in fact, no difference between the two branches of chemistry and opened up the possibility of producing synthetic organic chemicals (including polymers) from what were then cheap and abundant inorganic raw materials.

The first commercial production of organic chemicals was largely from wood, coal, and other naturally occurring organic materials. For example, methyl alcohol was obtained as a by-product from the production of charcoal, benzene and toluene were by-products from the production of coke, and ethyl alcohol was produced by the fermentation of molasses. In Germany, an extensive synthetic-organic chemical industry was based on calcium carbide and acetylene, both of which could be readily produced from coal and limestone.

In the United States, development of the synthetic-organic chemical industry followed a different path. Since we were becoming more and more dependent on petroleum to satisfy our energy needs, refineries were producing large quantities of off-gases with little value except as low-grade fuel. American industry soon found ways to convert these cheap off-gases to valuable chemical products. In 1920, Standard Oil of New Jersey (now Exxon) and Carbide and Carbon Chemical Corporation started producing isopropyl alcohol by absorbing propylene gas in sulfuric acid. A large synthetic chemical industry began to develop around the use of these refinery by-products, and *petrochemicals* (synthetic chemicals produced from petroleum) were soon being produced in ever-increasing quantities.

Polymers

By definition, a *polymer* is simply a large or giant molecule built up by the repetition of hundreds or thousands of small simple chemical units. Some polymers are produced in the form of long chains; others have a complicated three-dimensional structure. Naturally occurring polymers (such as wool, cotton, silk, jute, rubber, and cellulose) have been important throughout the history of mankind. Produced by living organisms, these *natural polymers* are dependent on agriculture, forestry, and animal husbandry.

In 1839, American inventor Charles Goodyear developed a process for *vulcanizing* rubber which made possible the automobile tire and many other uses of rubber. As it comes from the tree, rubber is a highly unsaturated natural polymer with the consistency of soft toffee. In the process of vulcanization, sulfur is added to remove unsaturation and thereby create a *modified polymer* which is harder and more suitable for industrial use. By 1870, another modified polymer, cellulose nitrate or Celluloid®, had come into common use for combs, Kewpie dolls, and other small objects—and not long afterward made possible the motion picture film. The first purely synthetic polymer, however, was the phenol-formaldehyde class of resins produced in Germany in 1907 and still of moderate importance under trade names such as Bakelite®. By the 1930s, the Germans had also developed polystyrene, polyvinyl chloride, and the synthetic rubbers; in the late 1930s, polyethylene was produced by the English at about the same time that Du Pont was introducing nylon in the United States. The polymer binge was beginning.

Synthetic Polymers in the United States

The end of World War II saw very rapid growth of the synthetic polymer industry in this country, largely the result of three factors: (1) Natural polymers, such as wool and cotton, have always been dependent on agricultural conditions and sometimes on political climates in different regions of the world; because of this, their cost and availability has been subject to very wide fluctuations. (2) The raw materials required for synthetic polymers were becoming cheap and easily available through our "abundant" supplies of petroleum and natural gas. (3) Different types of polymers were being introduced, based on both German and American technology; by proper choice of raw materials, it was soon found that the properties of a synthetic polymer could be varied to meet a particular need—it was becoming literally possible to custom-design a polymer for almost any given application.

In 1950, total United States production of all synthetic organic chemicals (including polymers and their raw materials) was 20 million tons. By 1960, this had more than doubled to over 48 million tons. By 1970, it had more than doubled again to 117 million tons, and, by 1979, it had reached the staggering figure of 177 million tons—a value greater than that of our entire production of metals, including steel, aluminum, and copper. United States production of polyethylene (which includes garbage bags and freezer bags) is now greater than our production of aluminum; production of polystyrene, polypropylene, and the polyvinyls are all greater than our production of copper, lead, or zinc. Each year we produce almost 3 million tons of synthetic rubber and more than 2 million tons of synthetic detergents, equivalent to over 20 pounds of detergent for each man, woman, and child in the country. We produce some 700,000 tons of synthetic organic pesticides, equivalent to over 6 pounds for each inhabitant and seemingly enough to poison every insect in the entire United States (Chapters 3 and 6 deal further with the pesticide story).

To illustrate the complexity of today's synthetic organic chemical industry, Figure 22 shows how one refinery product, *ethylene,* can serve as the parent for an entire family of different products. Off-gases produced by conventional refinery operations are sent to cracking furnaces to produce ethylene gas, and the ethylene is separated, purified, and used as a chemical raw material for a host of synthetic products. Note particularly the complex interrelationship where one product becomes parent for yet another product. For example, ethylene is combined with benzene to form the product ethylbenzene; this in turn is converted to styrene, which can then be polymerized to either form polystyrene or (in conjunction with butadiene) used to make synthetic rubber.

In addition to the family of synthetic organic chemicals based on ethylene, there are three other important organic-chemical raw materials that serve as parents for different families of products.

1. ***Propylene.*** This material, largely obtained as a by-product of other refinery operations, is next in importance and serves as parent for such products as polypropylene, acetone, polyurethane, the acrylic fibers, the epoxy resins, and synthetic glycerol.

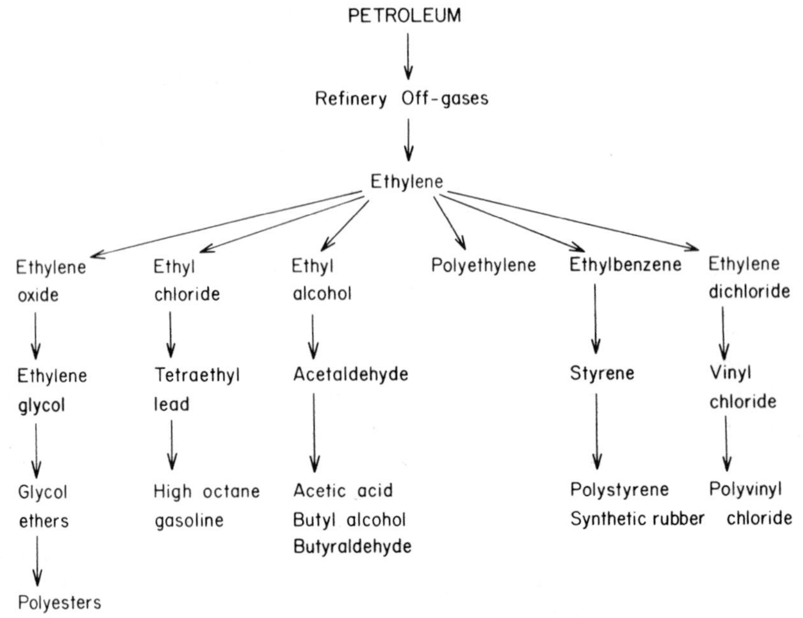

Figure 22. Chemical products from ethylene.

2. The Butanes. These serve as parents for butyl alcohol, methyl ethyl ketone, and nylon as well as several types of synthetic rubber.

3. Natural Gas. Mostly methane, this material serves as the source of practically all of the synthetic ammonia so vital to our fertilizer industry. In addition, natural gas is parent to most of our methyl alcohol, formaldehyde, and the entire series of phenol-formaldehyde and urea-formaldehyde polymers.

Perhaps a better way to point out the importance of these synthetic organic chemicals is to give a few examples from our daily life. By the late 1970s, they had taken over almost three-fourths of the apparel market. Sweaters are practically all acrylics, such as Orlon®. Women wear polyurethane foundation garments, nylon hose, dresses, and lingerie of nylon or polyester. Most women's handbags are now vinyl. Men's topcoats are normally wool-polyester or wool-nylon blends, and they wear cotton-polyester shirts (no ironing), cotton-polyester underwear, and business suits of cotton-polyester or wool-polyester blends. Raincoats are vinyl or treated polyester. The list is endless, but the fact is that cotton represents only 25 percent and wool only slightly over 1 percent of our clothing sales today.

Our houses are built using plywood bonded by a phenol-formaldehyde resin. The kitchen floor is finished with vinyl tile, and the basement floor with vinyl-asbestos tile. Electric wiring is insulated with a polyethylene or polyvinyl coating, and the outlets, wall plugs, and switches are of phenol-formaldehyde resin. The roof is insulated with polyurethane. The exterior and interior are painted using synthetic lacquers, enamels, and solvents. The bathrooms may have fixtures of Lucite® (polymethyl methacrylate) and drains of plastic. Nylon, acrylic, or polypropylene carpets most of our floors today, and our draperies are of nylon or polyester. We sleep on synthetic foam rubber mattresses, with cotton-polyester sheets, acrylic blankets, and Dacron® pillows. Most of our furniture is padded with polyurethane and covered with polyvinyl or nylon, our dining table may well have a urea-formaldehyde top, and almost all kitchen counters are covered with Formica® (a melamine-formaldehyde bonded resin).

The car we drive to work is powered by gasoline and lubricated with a synthetic engine oil. Its tires, belts, and hoses are all of synthetic rubber, the carburetor and distributor are plastic, and the engine is protected by synthetic ethylene-glycol antifreeze. Car seats are usually polyurethane or synthetic sponge rubber with vinyl covers, and windows are of safety glass with polybutyral laminant.

Most of the food we eat is grown using synthetic organic pesticides and synthetic ammonia fertilizers made from natural gas. It is processed using synthetic flavorings, colors, extenders, emulsifiers, and preservatives. It is packaged in polyethylene and stored in refrigerators which are insulated with polyurethane and cooled by a synthetic refrigerant such as the Freons (chlorofluorocarbons). We may satisfy our thirst with synthetically flavored and sweetened "diet" drinks, and we ease our aches and pains with synthetic pain-killers, sleeping aids, and "happy pills" derived from petrochemicals. We wash our clothing, dishes, floors, windows, and even ourselves with synthetic detergents; we water our gardens with plastic hoses and fertilize our lawns with synthetic fertilizers and weed killers. Our children play with plastic games and toys. We exercise and entertain ourselves with synthetically bonded fiberglass skis, tennis rackets, campers, and sailboats. For the last act of all, many of us are now even being buried in synthetically bonded fiberglass coffins. It is no exaggeration to say that synthetic organic chemicals rule our lives from the womb to the tomb!

What Do We Do Next?

Prior to World War II, practically everything we had came from natural products. We slept on hair mattresses with cotton sheets and woolen blankets. Our clothing was usually wool or cotton and occasionally silk. We used to walk to school or work in leather shoes; if it rained, we wore rubber overshoes and a rubber raincoat. We washed ourselves with soap made from natural fats and oils. Our food consisted of unrefined or partially refined natural products with no added sugar, synthetic vitamins, or other chemical additives. Milk was delivered in returnable glass bottles, and meat was cut to order by the butcher and wrapped in plain paper.

The development of synthetic organic chemicals and polymers has greatly expanded our material world and has freed us to a large extent from our former dependence on natural products which were subject to rapid fluctuations in price and availability. Now, however, we must face the fact that the price of all of our synthetic materials will follow the price of crude oil. While it is true that, at one time, our petrochemicals were produced largely from waste gases of very little value, today's petrochemical industry accounts for roughly 5 percent of our total use of petroleum. There is no way this demand can be met by the available waste gases. Manufacturers of petrochemicals are being forced to use raw materials such as propane and butane, materials that have ready markets of their own. Some manufacturers are even resorting to naphtha, itself a gasoline fraction, to produce the ethylene and propylene needed for petrochemical raw materials.

What we must also recognize at this point is that it is quite impossible to return to natural products for our needs. The wool, cotton, silk, fat, and other natural products available today would supply only a very small fraction of the present demand for clothing, soap, packaging materials, and the like. In addition, the price of the natural products now follows the price of the synthetics. For example, when the price of polyester goes up, an increased demand for cotton places additional pressure on supplies of cotton; the price rises, and soon cotton and polyester are again competitive in price.

As in the case of energy, *the one sure thing we can do is to minimize the waste of our synthetic organic chemicals and polymers.* A good place to begin is in the use of packaging materials, much of which could be eliminated without hardship. We can go back to refillable milk and soda bottles, we can save and reuse plastic bread wrappers, we can try to cut down the weekly pile of containers that every family must dispose of these days. Not only is our throw-away economy becoming more and more expensive, sooner or later we must realize that these materials are being bought only at the price of a reduction in the amount of energy available to us.

It is, of course, possible to produce synthetic organic chemicals from raw materials other than petroleum. Before World War II, coal served as the basis for much of the synthetic-organic chemical industry, and some people state casually that we could always go back to coal. The cost would be incredibly high, however, since it would require the changeover of an entire technology with investment of billions of dollars in new processing facilities. It is very unlikely that any serious steps will be taken in this direction while there are remaining supplies of petroleum and natural gas—that is, until we have no alternative except a return to coal.

Unfortunately, when it comes to our synthetic organic chemicals, which are now just as important as our energy and mineral resources, government activity has been limited pretty much to a yearly compilation of statistics. Essentially no help at all has been given to the synthetic-organic chemical industry. What government action there is is often strictly negative; environmental regulations, which may in themselves be desirable, are often applied in a very arbitrary and restrictive way to this important source of materials.

OUR MINERAL RESOURCES

We have seen that the occurrence of the fossil fuels is somewhat predictable, since it is known that they are found only in sedimentary rocks or basins. The occurrence of minerals is not as predictable. Mineral deposits were formed billions of years ago during the cooling of the outer shell of the earth and, in more recent times, as a result of volcanic activity that brought molten material from the earth's core to the surface. Some fortunate regions of the world, such as Canada, the Soviet Union, Australia, and South Africa, have abundant supplies of critical minerals; others, such as Japan and Great Britain, have very little. It is this unbalanced distribution of minerals that makes certain nations of the world today so dependent on imports.

In the United States, many of our basic raw materials, such as sand, gravel, clay, salt, sulfur, magnesium, and iron ore are found in almost unlimited quantities. Others, such as lead, zinc, gold, and platinum occur in modest quantities but can be conserved through recycling and, if necessary, turning to the lower-grade ores. Still others, such as asbestos, chromium, fluorine, and mercury occur in only very small amounts; for these, recycling will become increasingly essential, and suitable substitutes may have to be developed in many instances.

When we speak about the availability of minerals, we must make a distinction between the *known reserves* and the *theoretical abundance* of a given material. The theoretical abundance is simply the total amount of the particular mineral that is estimated to be present in the earth's *crust* (usually considered to be the outer mile of the earth's surface). These estimates of abundance are based on geological research and on analysis of samples taken during mining and drilling operations. Estimates of relative abundance at present involve considerable uncertainty, since much of the earth's surface has been inadequately studied. The radius of the earth is 4000 miles; the deepest mines go down only some 2½ miles and the deepest wells some 6 miles, so it is apparent that we have only scratched the surface.

Known reserves are deposits of minerals that have actually been discovered and quantified. Normally, these include only minerals that can be recovered economically under prevailing conditions; thus the concept of known reserves is flexible, and the figures may increase as prices rise or as technological improvements make possible the mining of lower-grade ores. Also, industrial companies have been known to deliberately understate their reserves to reduce taxes or minimize the possibility of expropriation by foreign governments.

If we compare the known reserves of important metals such as aluminum, iron, copper, and lead with the theoretical abundance of these materials in the crust of the earth, we find that the theoretical abundance is roughly a million times greater. It appears the total amount of these materials present in the earth is sufficient to supply our needs for millions of years at our present rate of consumption. This does not guarantee that we will have no problems, however, since much of the material may be distributed in such low concentrations that recovery may not prove to be economically feasible. In addition, the highly concentrated deposits are distributed without relation to population or demand. As the concentrated ores become further depleted, there is no question that the "have-not" nations will eventually face serious problems in maintaining access to many essential raw materials.

It is increasingly apparent that many of our important minerals occur naturally in two ways: (1) A relatively small number of rich deposits that can be economically exploited, and (2) a tremendous quantity of the material occurring as only a few parts per million and distributed throughout much of the rock that makes up the crust of the earth. Because of the tremendous amounts of energy required for their extraction, it is unlikely that much of this low concentration material can ever be exploited. The critical question is not so much the total quantity of a given material present in the earth's crust but rather how close we are to running out of the rich deposits.

Availability of Mineral Resources

Most of the mineral resources that are traded internationally move from the undeveloped to the developed nations. In addition, many of these are located in undeveloped nations whose economies are strongly or almost entirely dependent on one particular mineral (examples include copper in Chile, bauxite in Jamaica, and tin in Bolivia). For this reason, it is important to remember that, although price fluctuations can be inconvenient to developed nations, they can be absolutely disastrous to much of the undeveloped world.

From the start of the industrial revolution, the United States and western Europe had access to large areas of the world acquired by both exploration and conquest. As technological society became more and more complex, local supplies of raw materials became inadequate; industry became more and more dependent on materials imported from colonies and from the "new" parts of the world. Actually, it is unlikely that mineral exploitation was responsible for any significant amount of colonialism during the eighteenth and nineteenth centuries. Imports to England and the United States at that time were predominantly such items as wool, cotton, silk, timber, tobacco, grain, salt, sugar, rum, and, of course, slaves. When the twentieth century created a demand for rubber, oil, and minerals, however, these were at first supplied mainly by Europe's colonies.

Since the end of World War II, colonialism has for all practical purposes ceased to exist throughout the world. Not surprisingly, however, the undeveloped world considers much of the present-day commerce and investment by the developed world to be just the same old colonialism—only now with large international corporations, such as Exxon, playing the part of the "bad guy." Whether justified or not, this hangover from colonialism is something the developed nations will continue to face in their negotiations with many of our suppliers of essential raw materials.

Today it is obvious that the pendulum may have swung too far in the opposite direction. Many American corporations have invested large amounts of money and technology in developing foreign resources, only to have agreements broken and costly new facilities arbitrarily expropriated by the host government. The establishment of OPEC, an out-and-out monopoly to limit production and control prices, is a classic example of the undeveloped world very effectively using Western technology for its own interests. In view of our mutual needs, we must look forward to a time when differences will gradually be forgotten, and when agreements can be worked out that will yield prices fair to both the developed and the undeveloped nations.

First, Some Good News

Table 5 lists those minerals that occur in such large quantities that the United States has sufficient reserves of its own to last for at least 100 years at the present rate of consumption. Note particularly that this includes most of the raw materials used in our daily life. For example, we have almost unlimited amounts of *limestone, clay, sand, stone,* and *gravel* required for the production of portland cement and concrete. In Wyoming, we have extensive deposits of *soda ash* (sodium carbonate) for the manufacture of glass. We have unlimited amounts of *common salt* (sodium chloride) in sea water, salt lakes, and as deposits of rock salt; these can also be used as a source of *chlorine,* caustic soda (sodium hydroxide) and manufactured soda ash. Also in sea water, we have unlimited quantities of *magnesium.*

We have, in addition, substantial reserves of important metals such as *copper, gold, lead, molybdenum, silver, titanium,* and *vanadium.* We have virtually unlimited amounts of *sulfur,* occurring both as elemental sulfur and as sulfide ores such as pyrites (iron sulfide or fool's gold). These can be used for the manufacture of sulfuric acid, the "workhorse" of the chemical industry. We have substantial reserves of *potassium* and *phosphorus,* both of which are required for the production of agricultural fertilizers.

Our almost unlimited quantities of *iron ore* are widely distributed, with particularly important reserves in northern Michigan and Minnesota. It is true that the high-grade Mesabi deposits are starting to run out, and most of our iron ore is currently imported from Canada, Venezuela, and Brazil; however, we have tremendous deposits in northern Minnesota of a lower-grade iron ore (taconite), which can be *beneficiated* (concentrated to obtain higher purity) and used satisfactorily when and if needed.

Most of our *aluminum* is now produced from bauxite (aluminum oxide), which is imported from Guinea, Jamaica, and Surinam. While the United States has only limited reserves of lower-grade bauxite, we do have unlimited quantities of aluminum present as aluminum silicate clays, and these can be chemically processed to recover aluminum oxide, a satisfactory (although more expensive) replacement for the imported bauxite when needed.

The Bad News

Table 6 lists those minerals for which the United States is at least 50 percent dependent on foreign sources. It is a rather long list but not as grim as it might first appear. In several cases, we have our own modest domestic reserves but the foreign materials can currently be obtained at lower cost than the domestic materials. Several items could be recovered as by-products from the refining of other ores if prices were high enough to make recovery justified. For others, adequate substitutes could be developed if we were to be entirely cut off from foreign supplies.

The three most critical minerals on this list are chromium, manganese, and nickel. *Chromium* is an essential alloying element for all stainless steels and most alloys resistant to high temperatures. There is no known substitute. The aerospace, chemical, and petroleum industries are all heavily dependent on chromium alloys. The United States has small reserves of low-grade chromium ore, but these are not now being worked, and all our chromium is imported at present. The Union of South Africa and Zimbabwe (formerly Rhodesia) contain most of the world's known reserves, and it is probable that Soviet Russia also has substantial reserves. World demand for chromium has increased sharply in recent years.

Manganese is an essential alloying element for practically all ordinary steels,

Table 5. Minerals with Adequate United States Reserves

Material	Life of Known Reserves	Occurrence
Aluminum	Almost unlimited	Widely distributed, as aluminum silicate clays
Barium	250 years	Widespread deposits
Boron	100 years or more	California, Great Salt Lake
Bromine	Unlimited in sea water	Sea water, salt lakes
Calcium, lime, cement	Almost unlimited	Widely distributed, as limestone and gypsum
Chlorine	Unlimited in sea water	Sea water, rock salt
Clay	Unlimited	Widely distributed
Copper	220 years	Arizona, Utah, Montana, New Mexico
Feldspar	Unlimited	Widely distributed
Gold	100 years	Nevada, South Dakota
Iodine	180 years	Michigan, Oklahoma
Iron	Almost unlimited	Michigan, Minnesota
Lead	120 years	Missouri
Lithium	185 years	North Carolina, Nevada
Magnesium	Unlimited	Sea water
Molybdenum	300 years	Colorado, New Mexico
Phosphorus	200 years or more	Florida, North Carolina
Potassium	150 years	New Mexico (also tremendous reserves in Canada)
Sodium carbonate (soda ash)	Unlimited in sea water	Natural deposits of sodium carbonate in Wyoming
Sand, stone, gravel	Unlimited	Widely distributed
Silver	170 years	By-product from refining of copper, lead, zinc ores
Sulfur	Unlimited	Widely distributed
Titanium	1000 years	New York, Virginia
Vanadium	4300 years	Arkansas, by-product of uranium ores

Table 6. Minerals with Inadequate United States Reserves

Material	Present Dependence on Imports (%)	Source	World Reserves at Present Rate of Consumption (Years)
Antimony	50	South Africa, United Kingdom, France, China	73
Arsenic	90	Mexico, France, Sweden	250
Asbestos	86	Canada, South Africa	47
Bismuth	85	Peru, West Germany, United Kingdom	33
Cadmium	54	Australia, Canada, Mexico	1000
Chromium	100	South Africa, Zimbabwe	450
Cobalt	100	Zaire, Zambia	145
Columbium	100	Canada, Nigeria, Brazil	1000
Fluorine	85	Mexico, South Africa, Spain	150
Manganese	90	Brazil, Gabon, South Africa	300
Mercury	50	Algeria, Spain, Yugoslavia	90
Nickel	90	Canada, Norway, New Caledonia	150
Platinum	100	South Africa, Soviet Russia	200
Tantalum	100	Canada, Nigeria, Brazil	670
Tin	100	Malaysia, Thailand, Bolivia	160
Tungsten	55	Canada, Bolivia	140
Zinc	60	Canada, Honduras, Thailand	40

and—again—there is no satisfactory substitute. The United States has several large low-grade deposits of manganese ores which could be developed if needed. There are also adequate world reserves of high-grade ore in South Africa, Brazil, Gabon, and Soviet Russia—although the latter is not one of our sources. In addition, it is possible to recover manganese from ocean-floor nodules, and this may eventually become important to us.

Like chromium, *nickel* is an essential alloying element for stainless steel, high-temperature alloys, and corrosion-resistant alloys such as Monel metal. The United States has small reserves of high-grade ore and fairly extensive reserves of low-grade ore which could be used if necessary. There are substantial world reserves of nickel in Canada, Norway, New Caledonia, and Soviet Russia; it is also possible to recover nickel from ocean-floor nodules. World demand for nickel has increased substantially since World War II.

The following materials are obtained largely as by-products from the refining of other ores, and much of this potentially available material is now being discarded in the United States since it is less expensive to simply import foreign products.

1. ***Antimony.*** This material is chiefly used as a lead-antimony alloy for batteries. Most antimony is a by-product from the refining of lead and silver ores, and considerable amounts are recovered from recycled batteries. Introduction of other types of batteries has reduced the demand for antimony in recent years.

2. ***Arsenic.*** This material is primarily used for the manufacture of arsenical pesticides. Most arsenic is recovered from flue dust originating in the refining of gold and copper ores. Demand for arsenic has fallen in recent years with the widespread use of organic pesticides.

3. ***Bismuth.*** This material is largely used in the manufacture of pharmaceuticals and low-melting alloys. Bismuth is recovered as a by-product from the refining of lead ores.

4. ***Cadmium.*** This material is largely used for electroplating, pigments, and special alloys. Cadmium is recovered as a by-product from the refining of zinc ores.

5. ***Cobalt.*** This material is essential for the production of magnets, cutting tools, and special high temperature alloys for the aerospace industry. It is recovered as a by-product from the refining of copper ores.

Other critical minerals include the following.

1. Asbestos. This material is chiefly used for the manufacture of vinyl-asbestos floor tiles and asbestos-cement pipe. The United States has substantial reserves of asbestos, but most of it is now being imported. Since asbestos has been shown to be a dangerous carcinogen, it is apparent that several uses of this material (such as for insulation or ceiling tiles) will gradually disappear.

2. Columbium. This material is an alloying element in the manufacture of stainless steel, and also finds some use in special alloys for metalworking, transportation, and the oil and gas industries. The United States has very small known reserves, but there are ample reserves in Canada, Brazil, and Soviet Russia.

3. Fluorine. This material is required for the production of aluminum and is also used as a flux to remove impurities in the production of iron and steel. In addition, fluorine is widely used in the chemical industry for the production of aerosols and refrigerants. Although most of this material is now imported, the United States does have substantial reserves, and it is also possible to recover fluorine as a by-product from the manufacture of phosphate fertilizer. Use of fluorine has grown rapidly since World War II. This growth is probably coming to an end, with most use of fluorine in aerosols now banned because of possible adverse effects on the world's ozone layer.

4. Mercury. This material's two chief uses are in the electrical industry (mercury switches and mercury vapor lamps) and in the chemical industry (production of caustic soda and chlorine). The United States has modest reserves, and there are also modest reserves in Spain, Yugoslavia, and the Soviet Union. Both the price of mercury and the demand for mercury have shown wide fluctuations in recent years, with our domestic production becoming significant only when the price is right.

5. Platinum. This material is widely used as a catalyst in the chemical industry and in the refining of petroleum. There has been a substantial increase in the demand for use in automobile emission controls. The United States has substantial platinum reserves but only very small production as a by-product from the refining of copper ores. Demand has increased sharply in recent years, with most of our imports coming from the substantial reserves in South Africa and Soviet Russia.

6. Tantalum. This material is a moderately important alloying element for the manufacture of electronic components and metalworking machinery. While the United States has very small reserves and no current production, there are substantial reserves in Zaire, Canada, and Soviet Russia.

7. Tin. This material is widely used for the manufacture of cans and containers (tinplate), low-melting solder, and alloys such as brass and bronze. Because it is expensive and because there are adequate substitutes for many of its uses, the use of tin has been roughly constant for many years. The United States has very small reserves of its own, but there are widespread foreign reserves in South America, the Far East, and Soviet Russia.

8. Tungsten. The chief use of this material is for cutting and wear-resistant tools. With limited reserves in the United States, most of our tungsten is imported. There are moderate reserves in Canada, China, and the Soviet Union.

9. Zinc. This material is widely used in galvanizing, special alloys, paints, chemicals, and rubber products. There has been little growth in consumption during recent years because of the availability of substitutes. Even though the United States has substantial reserves, some 60 percent is imported, and there are also modest world reserves in Canada, Australia, and the Soviet Union.

Future Availability of Minerals

The problems involved in maintaining adequate supplies of mineral raw materials will be small in comparison to our problems with energy. The most annoying problem will be the increasing need to draw on lower-grade ores. For example, to produce 1 pound of copper now takes twice the amount of ore that it required about thirty years ago. This has increased the amount of ore that must be processed, the amount of energy required for processing and its cost, and the capital required to finance mineral exploitation. With these factors added to the eventual cost of a product, it is no wonder that it is often cheaper to import high-grade foreign ores than to develop lower-grade domestic deposits.

Environmental problems have become enormous and frightfully expensive. Beneficiation of ores such as taconite produces huge quantities of potentially harmful waste material which must be disposed of safely. About 20 percent of the capital expenditures of the steel industry now go to bring existing plants into compliance with new environmental laws. About one-third of the cost of a new copper smelter is for equipment to control pollution by solid and gaseous effluents.

For many years, the reasonably low cost of producing and processing minerals was not reflected in their price. Solid and gaseous effluents such as sulfur dioxide were simply sent up the plant smokestack, liquid effluents were dumped into the nearest river, and enormous piles of waste material were left to spoil the landscape. Today, of course, these "quality of life" factors are gradually being included and will reflect a substantial increase in the eventual cost to the consumer.

Another potential problem lies in our dependence on imports from unstable regions of the world. This has already affected our supply of cobalt from Zaire and of chromium from Zimbabwe. It is certainly not reassuring to be in a position where, for example, civil war in South Africa might cut off the supply of several of our critical raw materials.

To minimize the effects of such possible interruptions to our supply of critical raw materials, the federal government has for some years promoted a policy of stockpiling roughly a two-year supply of such materials. This has involved uncertainties of its own, and mistakes have been made; a stockpile of quartz crystals, for example, suddenly became obsolete when synthetic substitutes were developed for use in electronic equipment. Considering the world situation today, however, stockpiling appears to be a sensible precaution and well worth its modest cost.

Politics and Cartels

The question of insuring future availability of our critical raw materials is a good illustration of the complex interrelationship among economics, sociology, technology, and political science. It is no exaggeration to say that Middle East oil now dictates our entire political posture toward that part of the world. The 1973–74 Arab oil embargo was largely triggered by United States support of Israel during the Yom Kippur War. The more militant Arab nations continue to try to use the "oil weapon" to force further concessions from Israel and a solution to the PLO question. Settlement of the Iranian hostage crisis was clearly hindered by the desire to maintain free access to Persian Gulf oil.

A similar situation exists to a smaller extent in our relationship with several of our mineral suppliers. It is no coincidence that some of these nations have been generously provided with American foreign aid. When Zimbabwe-Rhodesia was undergoing internal strife, United States official policy was considerably influenced by a desire to maintain access to their supplies of chromium. Similarly, our criticism of racial policy in South Africa is tempered by the fact that they furnish many of our most critical minerals.

With the amazing success of OPEC, it is not surprising that attempts have been made to control the production and price of other raw materials in similar fashion. Actually, there is nothing new about the formation of an association or *cartel* to control production and maximize profits. The first cartels involved a "gentleman's agreement" between private manufacturers, but even these often involved some degree of official or unofficial government participation. During the nineteenth century, oil, whiskey, and sugar "trusts" were established to control production and set minimum prices. In Germany, government-sanctioned cartels were widely used to monopolize production of critical materials; for example, I. G. Farben controlled world production of dyes and synthetic organic chemicals for some thirty years until their monopoly was ended by World War I.

In the meantime, the Sherman Antitrust Act had been passed by the United States Congress in 1890, and President Theodore Roosevelt used his "big stick" to break up many of the trusts in the early twentieth century. The Federal Trade Commission (FTC) was established by the Clayton Antitrust Act of 1914, but none of this had any effect on the international cartels.

The early 1920s saw a scheme to control the price of rubber, and the Inter-American Coffee Agreement established in 1940, with the United States government arranging a quota system for coffee exports from Latin America. Similar agreements have also been established over the years for sugar, wheat, and olive oil.

In 1967, Chile, Peru, Zaire, and Zambia formed an alliance of copper producers to control production and prices. Increased consumption of copper during the early 1970s caused a rapid increase in the price, which peaked at over $1.50 per pound in April of 1974. A year later, however, the price had dropped to slightly over 50 cents per pound; despite several cuts in production by the copper alliance, the price has shown no sharp increase in the past few years.

What was behind this failure of the copper cartel? Since the United States has substantial reserves of copper, much production can be achieved using domestic resources. The chief use of copper is in the electrical industry where microwave towers and communication satellites are now replacing copper transmission lines. There is less demand for copper alloys such as brass and bronze, and plastic pipe has replaced copper water and drain pipe in most building construction. Finally, if the price of copper were to rise too much, aluminum could be substituted for most of its uses.

In 1974, the chief world producers of bauxite established the International Bauxite Association (IBA). Their purpose was not to control prices and production but to establish more bauxite-processing facilities in the producing nations. For years, it had been customary for the aluminum companies (such as Alcoa and Reynolds) to ship bauxite to the United States where they would chemically concentrate the ore from about 26 to 48 percent aluminum oxide and then electrically reduce the aluminum oxide to aluminum metal using a molten bath of cryolite (sodium aluminum fluoride). What the bauxite producers wanted was to have at least the chemical processing facilities and, perhaps, some of the aluminum production facilities located in their countries to improve the local economy.

Although the bauxite cartel has hardly been an overwhelming success, there has been some yielding to the bauxite producers; it is obvious, however, that if the cartel goes too far, the big American aluminum producers will simply shift to domestic clays for their source of aluminum oxide. There have also been some motions made toward establishment of cartels for tungsten, silver, tin, mercury, iron ore, and natural rubber, but in general, these have been no more successful than the efforts of the copper and bauxite producers.

Chapter Six

Pollution and Public Health

Primitive man had no pollution problems. His very simple needs were supplied by natural products: pure water in every river and stream, food and clothing from local vegetation and animals, and ample firewood for the taking. A 1-minute walk into the woods took care of his sanitary needs. When he died, his body was planted in the earth and recycled back to nature. A few stone tools, perhaps a clay pot or two, were soon the only evidence that he had even existed.

The Neolithic revolution and the development of cities brought a need for sanitation, particularly for a good water supply and a means of removing sewage. Archeological evidence shows that these problems were actually being solved more than 4000 years ago. Excavations of early civilizations in India, Crete, Greece, Egypt, and South America have turned up the remains of large conduits used to bring water to the cities, together with bathrooms, drains, and sewers to remove effluent. These, of course, were nothing in comparison to the elaborate systems developed by the ancient Romans. At its height, Rome had thirteen aqueducts bringing tens of millions of gallons of water into the city each day to supply fountains, baths, and other public facilities (a private supply of water could be obtained only by imperial grant). Many of the Roman baths were extremely impressive; the Baths of Diocletian, for example, covered several acres and could accommodate more than 3000 people at one time.

Rome also had some 150 public latrines and a system of drains running under the streets to carry off surface water and sewage. The great sewer of Rome, known as the *Cloaca Maxima,* was about 10 feet wide and 12 feet high at its point of discharge into the Tiber River.

Further growth of cities during medieval and renaissance periods found life becoming more complicated for the inhabitants. Special officials were appointed to see that water required for drinking and cooking was not polluted. Citizens were told not to throw refuse or dead animals into the water supply; tanners were forbidden to wash skins, and dyers were forbidden to dispose of their dye residues in the water, and even the washing of clothes was sometimes prohibited, causing great inconvenience. Street cleaning and garbage disposal became problems. Chamber pots were simply dumped out the window, and dead animals were often thrown into the street. With many city dwellers keeping horses, cows, and large numbers of hogs, geese, and ducks, the stench became so bad that several German cities passed regulations to forbid the construction of hog pens facing the street!

Figure 23. Weeds grow around boarded-up homes along Love Canal. The "Dead End" sign is particularly appropriate. (United Press International)

While public baths were available in the larger cities in those days, few people took advantage of them; most of the inhabitants of northern Europe seldom or never took baths, and it is little wonder there was always a good market for strong perfumes "from the East." The poorer people often had only one outfit of clothing, which they wore continuously, and the generally cold climate of northern Europe made even those who could afford it reluctant to change their clothing. Under these conditions, it is not surprising that parasitic insects were very common. Fleas were everywhere—in the ground, on domestic animals, on rats, and on most people as well. Today we know that rat fleas were responsible for the almost incredible spread of bubonic plague (the Black Death) in the fourteenth century.

Lice, too, were very common in these days and often referred to as the "gentleman's companion." Hairdressers kept a bottle of alcohol handy to use on customers showing evidence of the head louse. As late as the middle of the nineteenth century, there were over a million cases of typhus on record in England—all spread by body lice. Bedbugs became more prevalent during the nineteenth century because of the richer furnishings of Regency and Victorian homes. Evidently no consideration was shown by them in selecting a suitable host; during the reign of Queen Victoria, Tiffin and Son could proudly announce that they had received a Royal Appointment as "Bug-Destroyers to Her Majesty."

By the start of the twentieth century, however, improved public and personal sanitation had just about solved problems of bacterial contamination of air, food, and water. Filtration and chlorination were gradually introduced to purify water supplies; refrigeration and canning helped in the safe distribution of foods; better sanitation eliminated most transmission of disease by insects. Chemical contaminants were to become the major pollution problem of the twentieth century.

WHAT IS POLLUTION?

One of the most difficult aspects of the problem of pollution has been to develop guidelines defining what actually constitutes pollution. The rock concert that is such a joy to your teenager's ears is simply loud and unpleasant noise to the neighbors. Some people would never under any circumstances live in New York City while others would never live anywhere else. To the fisherman, anything that alters his favorite stream is pollution; to the industrial worker, the most important thing is a job—even if it does bring some contamination to the environment.

In the strictest sense, pollution can be defined as anything that in any way changes the natural environment. With such a definition, human life itself constitutes pollution. We make noise, we produce heat which must be dissipated to the atmosphere, we carry out agriculture, we consume food crops, and we produce waste products which pollute the environment. Of course, most people do not feel that their lives represent pollution; in fact, most of us take a rather opposite and selfish attitude. Normally, we consider pollution to be any change in the environment that we personally find objectionable; changes that give us pleasure are considered desirable.

In the 1980s, a commonly accepted definition of *pollution* is "an undesirable change in the physical, chemical, or biological characteristics of our air, land, or water that may or will harmfully affect human life or that of our desirable species. . . ." This really does not help us because it leaves open the question of *how much* change we can tolerate in our environment before it becomes "undesirable." Also, there will always be the question of what is or is not a desirable species. For example, some years ago when sprays were first used in an attempt to control the famous New Jersey mosquitoes, there was opposition from certain religious groups who felt that "God created the mosquito."

Since any human activity must, of necessity, cause some pollution, it is meaningless to talk about a zero level of pollution; the goal of any pollution-control program must be to determine an *optimal level* of pollution—that is, the point where the benefits to society of the proposed activity outweigh the harmful effects of the inevitable pollution that results.

In establishing optimal levels of pollution, it is important to remember that pollution control is extremely expensive and that its cost always increases sharply as the level of the particular pollutant is reduced. It is estimated that some 2 to 2.5 percent of our gross national product is now spent on pollution control programs, and in some mining and metallurgical industries the cost may be as much as one-third of the total production cost. In some cases, it might be a fairly easy matter to reduce an effluent concentration by 90 percent, but further reduction (say, 98 percent) would necessitate the spending of five or ten times as much for control. Since the cost of pollution control is eventually passed down to the consumers and taxpayers, it is essential that realistic goals be established and reevaluated from time to time to ensure that the benefits do, in fact, justify the additional expenditures.

One thing is quite clear: establishment of reasonable and acceptable pollution levels—that is, optimal levels—is extremely complicated and requires a good understanding of many environmental, economic, political, and technical matters. Unfortunately, there is almost always a battle between extremists, usually industrialists who want levels as high as possible and environmentalists who insist on zero pollution, with politicians just trying their best to please their constituencies. Since our modern technological society cannot exist without a certain degree of pollution, we obviously have two choices: (1) we set reasonable levels of pollution that industry can live with, or (2) we continue to pay more and more for less and less of the things that we use in our daily life.

Types of Pollution

It is possible to categorize pollution in several different ways, but there is always considerable overlap in any classification. In some instances, pollution is classified according to the *receptor* of the pollutant (such as air, water, land, and environmental pollution); in other instances, pollution is classified as either *degradable* or *non-degradable* (depending on whether or not the particular pollutant can be easily broken down when discharged into the environment); in still other instances, it is classified according to the nature of the *pollutant* (thermal, sulfur, and DDT pollution, and so on).

BIOLOGICAL POLLUTION

The primary problem of providing uncontaminated food and drinking water has been solved in the United States. Typhoid fever, cholera, and other diseases transmitted by contaminated food or water are nonexistent today. Each year there are a few deaths from botulism, usually from home-canned foods—almost never from commercial products. There are occasional incidents of ptomaine poisoning from improperly prepared or improperly refrigerated foods, and when we are away from home we all occasionally get the "traveler's complaint." On the whole, however, when we consider the enormous quantity of food that is produced, marketed, and consumed in the United States, there is no question that the food industry (with the encouragement of our pure food and drug laws) does an excellent job of protecting the public.

Another common type of biological pollution comes from the disposal of wastes such as human sewage, animal manures, agricultural wastes, and the refuse from slaughter houses and poultry-processing plants. These wastes all have one common advantage—they are readily degradable by bacteria naturally present in the environment, which literally feed on them and soon convert them back to carbon dioxide and water. Although these wastes often have a foul smell and are quite offensive, they do not present the environmental hazards of the chemical wastes. The only real problem with animal and agricultural wastes comes when the environment is so overloaded that they cannot be disposed of by the usual bacterial action.

At least from an engineering standpoint, the problem of sewage disposal has also been solved. It is true that many communities still dump untreated sewage into the environment, but this is merely because they do not want to spend money for treatment facilities. The most commonly used method of treating sewage wastes is the *activated-sludge process,* whereby the sewage is mixed with a recycled sludge and blown with air to biologically consume the organic material present in the sewage. The treated sewage is then allowed to settle, and the clear liquid is discharged to a river or lake. Usually the effluent leaving the sewage treatment plant is cleaner than the body of water into which it is dumped.

The substantial quantities of sludge produced by such a sewage disposal plant can sometimes present a disposal problem. Some communities merely dump the sludge into the ocean or in land-fill areas; however, this is wasteful since the sludge does have value as a fertilizer. Direct use of it as fertilizer has been studied but has run into opposition because the sludge is loaded with bacteria, particularly of the *coliform* type. In Milwaukee, the largest city in Wisconsin, this problem has been solved in a very constructive way by *calcining* the sludge (that is, heating it in a rotary furnace to a high temperature that dries and kills the bacteria) and then selling the product as Milorganite fertilizer.

Eutrophication

A more serious problem has developed in recent years because of the presence of nitrogen and, particularly, phosphorus compounds in sewage effluent. This is largely the result of detergent manufacturers' adding substantial quantities of phosphorus to their products (it permits them to reduce the amount of more expensive synthetic organic detergent present, and it does obviously improve the cleaning ability). Ordinary sewage treatment processes do not remove phosphorus; when this element enters a body of water, it acts as a fertilizer to the algae and other aquatic plant life present and greatly increases their rate of growth. This nutrient enrichment of water is known as *eutrophication;* it can be either natural or artificial and is always present to some extent in any lake or stream.

Eutrophication only becomes a problem when the phosphorus and nitrogen nutrients overload the natural processes in a body of water. Excess growth causes the water to become more turbid; solar radiation is more easily absorbed so that the water temperature rises, and this speeds up the biological processes. There is a prolific growth of moss and aquatic weeds; local fish species change from valuable game fish to rougher types such as carp. The lake may eventually become so loaded with decaying algae and aquatic vegetation (which remove dissolved oxygen from the water and cause fish kills) that it becomes a dead swamp.

One of the most publicized examples of eutrophication is, of course, Lake Erie. Fairly shallow, it has received tremendous quantities of nutrients from the heavily populated and highly industrialized upper Midwest. Large areas of the lake have become densely choked with weeds; mats of algae over 1 foot thick and covering hundreds of square miles have been reported. Rooted vegetation grows in thickly from the lake's shore. A slime forms on the beaches and offensive odors result from the dying and rotting algae. Commercial fishing has been sharply reduced, recreational use of the lake has declined, and lake-shore property has suffered in value. Lake Erie is not quite dead, but it is certainly not in good health.

In an attempt to reduce the amount of phosphorus released to the environment, several states (such as Indiana, Michigan, Minnesota, New York, Vermont, and Wisconsin—four of them bordering the Great Lakes) have banned the sale of all phosphorus-containing detergents. Unfortunately, there is no really satisfactory substitute for phosphorus in detergents. It was thought at one time that nitrilotriacetate (NTA) could be used in its place—as is currently being done in Canada—but the United States government feels there is a possibility that NTA may be a carcinogen, and our manufacturers are understandably reluctant to try it. Most manufacturers in this country are simply increasing the amount of soda ash, which is present to some extent in all detergent products.

The disposal of animal and agricultural wastes has not received much publicity. Food-processing plants, slaughterhouses, and poultry-processors usually discharge their refuse to the local sewage disposal plant, although they often require at least some preliminary treatment before they can be accepted by conventional facilities. Agricultural wastes such as chaff, cornstalks, and the like are seldom a problem and can be returned to the land to enrich the soil. By far the most serious challenge of agricultural wastes is the runoff from fertilized agricultural land.

The United States uses some 50 million tons of artificial fertilizer annually, including 10 million tons of nitrogen and 2 million tons of phosphorus. Most of this fertilizer is readily soluble in water, so that rainfall leaches these compounds from the land and carries them into our lakes and rivers. Crude estimates indicate that the runoff from fertilized land contributes at least half of the total human-generated nitrogen and perhaps 30 percent of the total human-generated phosphorus that enter our surface waters. Since nobody is seriously suggesting that we stop the use of artificial fertilizers, it is apparent that even complete banning of phosphorus in detergents would have little effect on water quality in the agricultural regions of our country.

The point that must be made about eutrophication is that *once it has gone beyond a certain point, it becomes almost irreversible.* For example, Lake Erie has a thick layer of nutrient-rich mud deposited over its entire shallow western end today. Even if all discharge of nutrients to the lake could be stopped, the material already present would prevent natural recovery of the lake for tens or even hundreds of years. To really clean up Lake Erie at this point would require dredging hundreds of square miles of lake bottom, an extremely difficult and expensive undertaking and one that would represent no real solution as long as large amounts of nutrients continue to be discharged into the lake.

Another problem from agricultural runoff has been the contamination of local farm wells and ponds with nitrates leached from fertilizer. Actually, the nitrates themselves are not particularly dangerous, but they can be reduced to toxic nitrites by bacteria present in the digestive tract. This is a particular hazard for infants; illness and infant deaths have been reported in agricultural regions such as southern Illinois and the Central Valley of California, where mothers are using more and more distilled or bottled water for their babies.

ENVIRONMENTAL POLLUTION

Several forms of pollution (such as noise, thermal, ultraviolet radiation, electromagnetic radiation, and microwaves) involve only a physical change in the environment. They are far more difficult to evaluate because they tend to be localized, are often impossible to measure quantitatively, and there is often considerable doubt whether or not they do actually constitute a hazard. For these reasons, their possible environmental impact is sometimes simply ignored.

Consider the case of noise pollution. At one time, the hustle and bustle of a big city was one of its chief attractions; today most cities present an incredible bedlam of roaring traffic, endless construction work, and continuous overhead air traffic as well. Nor are the suburbs necessarily more peaceful if one lives near an airport or interstate highway. There is no good evidence to indicate that this ordinary city hubbub causes physical harm, although prolonged exposure to noise of about the level of heavy road traffic can eventually impair hearing. The fact is that, when so many of our young people flock to lounges and discos where they are exposed to flashing lights and loud amplified music, it is impossible to convince anyone that even the loudest traffic noise can present a problem.

A much more serious aspect of noise pollution is found in the working place. It has been estimated that millions of Americans are threatened with hearing loss today because of noisy working conditions. There is also the question of how many accidents on the job occur because of noise levels, to what degree stress and fatigue are accentuated by noise, and whether noise can be held directly responsible for hypertension, ulcers, and even mental illness. Noisy working conditions are now fortunately being controlled through our labor laws, and environmental impact studies require an evaluation of possible pollution by noise in more and more areas.

The possible environmental effects of magnetic fields have surfaced in the last few years because of increased construction of high-voltage transmission lines. Electric utilities are transmitting electricity for greater distances at voltages ranging up to 1 million volts. Those living near such transmission lines have expressed some concern whether there are any adverse physical effects possible from exposure to these magnetic and electric fields. While animal experiments have consistently indicated that there is no hazard, it is easy to understand their concern—particularly when the lines start hissing, crackling, and emitting flashes of light on a dark and damp evening!

Microwaves, too, are being questioned as a possible health hazard in view of their increasing importance in our daily life. They are used to transmit telephone calls and television signals, for fast-cooking ovens, and not too long ago it was disclosed that the Soviets were using powerful beams of microwaves to monitor our communications at the United States embassy in Moscow. One of the "way out" ideas for utilizing solar energy mentioned in Chapter 4 involves the transmission of energy from space satellites, using broad beams of microwaves.

Apparently the microwave beams used in telephone and television transmission are of such low intensity as to present no hazard; however, restrictions have been placed by the federal government on microwave emissions from home ovens in response to a growing concern by the public. In view of the lack of information available regarding possible effects of exposure to microwaves, it is not surprising that these questions are being raised.

Polluting Our Global Environment

Another question frequently asked today concerns whether or not our increased scale of technological activity can adversely affect the entire global environment. We know, for example, that changes in the amount of carbon dioxide and particulate matter in the atmosphere can affect plant growth and rainfall. There is evidence that release of chemicals such as the chlorofluoromethanes can affect the earth's ozone layer, thereby increasing ultraviolet radiation and the incidence of skin cancer. It is also well known that even a slight decrease in global temperature could trigger another ice age; similarly, a slight increase in global temperature could cause considerable melting of the polar ice caps, bringing a rise in the level of our oceans and flooding most coastal cities. These are some of the known hazards, and there may well be other effects that have not yet been recognized or whose consequences we cannot yet determine.

In general, the question of changes in our global environment is extremely complex, and there are no easy answers. For example, consider the temperature of the earth. It is known that earth's climate has changed many times in the past and that the last great ice age, when much of North America, Scandinavia, and Asia were covered with great sheets of ice, retreated some 10,000 years ago. Since that time, the earth has experienced periods of warming and cooling where the average temperature changed by one or two degrees over a period of perhaps a century. A cooling trend in Europe during the seventeenth century produced such hardship that it has been referred to as the "little ice age." The first half of the twentieth century saw a pronounced warming trend that continued until about 1940. Since then, we have been back on a cooling trend, and there is no way to tell whether or how long this will continue.

These natural periodic fluctuations in temperature make it impossible to determine whether or not human activities have had any effect on average global temperature up to this point. The most disturbing aspect, however, is the likelihood that human-induced changes to global environment could well be irreversible and that, by the time a particular hazard is recognized, it might already be too late.

Global Temperature

The average temperature of the earth obviously depends on a balance between the amount of energy contributed by sunlight and the amount of energy radiated back to space. Sunlight reaching the upper atmosphere of the earth consists of ultraviolet through visible and infrared light. (This is sometimes referred to as the *spectral distribution* of sunlight.) Of unique importance in the upper atmosphere is the *ozone layer,* which normally absorbs practically all the ultraviolet radiation—a good thing for us, since we know this radiation causes skin cancer and is lethal to exposed microorganisms. Part of the sunlight that reaches our upper atmosphere is reflected back to space by clouds, ice crystals, and (to a smaller extent) dust; the rest passes downward through the earth's atmosphere and is further reduced in intensity by absorption in clouds, dust, and other materials. On a clear day, then, sunlight reaching the surface of the earth is about 10 percent ultraviolet, 45 percent visible, and 45 percent infrared light.

Of the sunlight striking the earth's surface, some will be absorbed and some will be reflected. As seen in Chapter 4, three-quarters of the earth is covered by oceans, and these will absorb practically all the sunlight that falls on them. Forests and croplands will also absorb most of the sunlight that reaches them. Ice and snow are good reflectors, but even the sunlight reflected from the surface of the earth has a good chance of being reabsorbed in our atmosphere before it can escape to outer space, so that a very high fraction of all sunlight reaching the earth's surface is converted to heat energy and warms the world in which we live.

Theoretically, human activity could change the world's energy balance in several ways. For example, much of the earth's surface has been quite drastically altered—cities and highways cover much of the landscape, forests have been converted to croplands, and deserts have developed and continue to expand in size. Has this been enough to significantly affect the amount of sunlight reflected or absorbed by the earth's surface? Fortunately, the global energy balance is determined primarily by the oceans and the polar regions. While changes in surface reflection and absorption resulting from human activity can strongly affect local conditions (cities tend to be pretty warm on a summer day), they are fortunately too small to change our overall global temperature.

A second factor which might influence global temperature is the tremendous amount of heat released to the environment through the burning of fossil fuels and the greater use of nuclear energy. This does produce local changes in climate, but on a global scale it is not as significant as might be expected. The total amount of heat generated by all human activity is about 1000 times smaller than the heat energy supplied by the sun. Since we know now that it is very doubtful that our consumption of uranium and the fossil fuels will ever be much higher than the present value, it does not appear that this source of heat will ever substantially affect our global temperature.

Another question often asked is whether the hundreds of millions of tons of dust, oil, smog, and acid fumes poured into the atmosphere each year by human activities have any significant influence on global climate. These materials do increase the cloudiness of our atmosphere and act in two ways to reduce global temperatures: they reflect a larger fraction of sunlight striking the upper atmosphere and they absorb a greater fraction of the sunlight passing down through the atmosphere to the surface of the earth.

There is no question that dust in the atmosphere causes a drop in global temperatures, as has been demonstrated vividly by volcanic eruptions. The best example of this is probably the eruption of Tambora in Indonesia in 1815, the greatest eruption in recent history. The following year, 1816, was cold all over the world and became known as "the year without a summer." In New England that year there was widespread snow between June 6 and June 11. Frosts occurred in every summer month; some crops did not ripen at all, and others rotted in the fields. The average summer temperature in Great Britain was 2 to 3 degrees centigrade below normal; all of northern Europe suffered, and there were serious food shortages in Ireland and Wales.

But, once again, the situation is not as serious as might be expected. When we consider the total "loading" of the atmosphere, it turns out that there is a natural injection of material into the atmosphere from forests (organic vapors), oceans (in the form of salt), deserts and other wind-blown dust, forest fires, and periodic volcanic eruptions. Over two-thirds of the pollution of our atmosphere is from natural sources, and less than one-third comes from human activities. Much of the air pollution by industry is now being reduced by filters, precipitators, and scrubbers. Unless we enter a period of violent volcanic activity, it seems unlikely that atmospheric dust will materially affect global temperature.

The Greenhouse Effect

To understand the highly publicized *greenhouse* effect,* now considered by some to be potentially the most serious consequence of human activity on global climate, we must return to the spectral distribution of sunlight. When the earth cools at night, its loss of heat to the atmosphere and eventually to outer space occurs in the infrared region of the spectrum. Simple gases (such as oxygen and nitrogen) do not absorb infrared radiation; more complex molecules, such as water vapor and carbon dioxide, do readily absorb and change it to heat energy (this is why the earth cools easily on a clear, dry night but does little cooling on a damp, cloudy night).

The greenhouse effect, according to its proponents, can be attributed to an increased concentration of carbon dioxide in the earth's atmosphere as a result of widespread burning of fossil fuels and increased agricultural activity (which speeds up the breakdown of biomass). The fact is that the earth does not lose quite as much heat at night as it used to, and there is a slight increase in global temperature. There is no question that global concentration of carbon dioxide in the atmosphere is now about 15 percent higher than it was prior to the industrial revolution, and it is estimated that it will be about 30 percent higher by the end of the twentieth century. Calculations show that—*other things being equal*—this increased carbon dioxide content of the atmosphere will cause about a 0.6 degree centigrade rise in global temperature by the end of the twentieth century.

Because of other factors that interact, it is very difficult to make a really accurate prediction of this kind. The oceans have about sixty times as much carbon dioxide as the atmosphere; in fact, the oceans readily absorb carbon dioxide gas from the atmosphere and tie it up as a loose bicarbonate solution in sea water. It is estimated that as much as one half of the carbon dioxide added to the atmosphere by human activity may have dissolved in the oceans; again, however, there is considerable question as to the accuracy of this figure and whether or not there has been time enough to establish an equilibrium between atmospheric carbon dioxide and that contained in the ocean depths.

A second factor which must be considered is that increased carbon dioxide content of the atmosphere causes an increased rate of photosynthesis. In other

*This is really a misnomer, since operation of a real greenhouse depends only partly on the radiation property of the glass and mainly on the use of a barrier to keep out the cold air.

words, plants grow faster in an atmosphere that is richer in carbon dioxide, and this serves to remove carbon dioxide more rapidly from the atmosphere and tie more up in the form of biomass.

A third factor is that increased global temperature means increased evaporation of water and, hence, increased cloudiness. This reduces the amount of sunlight reaching the earth's surface and the amount of heat lost by the earth during night-time cooling. Depending on the relative importance of these three factors, this could bring about either an increase or a decrease in global temperature!

Obviously, then, it is very hard to make any kind of precise evaluation of the importance of the greenhouse effect. It is even impossible to say whether or not it exists, since the maximum global temperature change, which could be attributed to it, is masked by the natural periodic global temperature variation during the past century or so. Since most of the carbon dioxide has been added in the past forty years, when global temperatures have been cooling, it might be argued that the greenhouse effect is nonexistent. While it now appears that the greenhouse effect is not something we have to worry about for the immediate future, it does bear close watching—particularly if we should get into another period of global warming.

CHEMICAL POLLUTION

The earliest recognition of the hazards of chemical pollution was centered around the diseases that could result from various occupations. Several writers of ancient Rome commented on diseases peculiar to sulfur workers and blacksmiths. They were particularly aware of the hard lot of the miners and often mentioned the pallor of the gold miner, "as pale as the gold that he tears out of the earth." The miners themselves recognized some of their hazards and frequently used bags, sacks, or membranes of some sort to cover their mouths to prevent excessive inhalation of dust.

During the Middle Ages and the Renaissance, there was further development of mines, salt works, foundries, glass works, tanneries, dye works, and soap plants, and several books were published describing the the occupational diseases of mine and smelter workers. In 1700, the Italian physician Bernardino Ramazzini published his classic *Discourse on the Diseases of Workers,* which served as the basis for occupational health study for over a century. Ramazzini described the diseases peculiar to workers such as miners, gilders, apothecaries, midwives, painters, wet nurses, and even cleaners of privies and cesspits. When it came to miners, he said, ". . . the mortality of those who dig minerals in mines is very great, and women who marry men of this sort marry again and again. . . ." In describing workers of flax, hemp, and silk, he said,

". . . a foul and poisonous dust flies out of these materials, enters the mouth, then the throat and lungs, makes the workmen cough incessantly, and by degrees brings on asthmatic troubles." Today we describe this as *brown lung disease.*

The industrial revolution brought a host of new problems, particularly those associated with the development of steam power and the expanded use of coal in the iron and steel industry. As early as the eighteenth century, coal smoke was blackening buildings in England. In 1775, Sir Percivall Pott, an English surgeon, published his *Chirurgical Observations,* in which he showed the high incidence of cancer of the scrotum among chimney sweeps was somehow related to their occupation, the first real evidence linking chemicals to cancer. Over 150 years later, British chemists found that 3,4-benzopyrene, a complex derivative of benzene and one of the chief constituents of coal tar, is indeed a very potent chemical carcinogen.

The chief health problem in those days was, of course, the terribly crowded condition of the cities. Nobody was paying much attention to smoke or chemical effluents. This was to change with the development of the Leblanc soda process at the end of the eighteenth century, the beginning of the modern chemical industry. The Napoleonic Wars had left France cut off from her supply of Spanish *barilla,* a seaweed that was burned to produce crude soda ash (sodium carbonate) for laundry purposes. The situation was desperate, and Nicholas Leblanc, a French chemist who was surgeon to the Duke of Orleans, had explored the possibility of producing soda ash from ordinary sea salt. After demonstrating the soundness of his process, Leblanc persuaded the Duke in 1791 to furnish 200,000 francs to exploit the idea. A factory was built, but the French Revolution shortly thereafter led to confiscation of all the Duke's property and the Duke lost his head. Leblanc was forced to reveal the secrets of his process to the Committee of Public Safety and saw his factory dismantled and his stocks of salt and soda sold. Never able to get started again and worn out by disappointment, he finally committed suicide in a madhouse in 1806.

The Leblanc process represented the first really large-scale chemical operation to meet consumer needs, but political conditions in France at the time were not favorable for industrial development and a high tax on salt greatly inhibited its use as a chemical raw material. When the English learned about the Leblanc process and when their tax on salt was reduced in 1823 from 15 to 2 shillings a bushel, very rapid growth of the industry followed, and for many years Great Britain's production of soda ash was greater than that of the rest of the world combined.

The first step in Leblanc's process was to heat salt with sulfuric acid to produce salt cake (sodium sulfate) with the evolution of hydrochloric acid gas. At first this gas was simply "sent into the air," but its harmful effects on vegetation and building materials soon became obvious, and the rapidly increasing production of soda ash and other chemicals created an intolerable nuisance. The government recognized that something would have to be done, and the first Alkali Act was passed in 1863, limiting the escape of hydrochloric acid gas to 5 percent of the total produced. Manufacturers installed towers to "condense" the gas by absorption in water, and it was found that the recovered by-product acid not only had a ready market in itself but could also be used as a raw material for the production of chlorine and bleaching agents.

During the last half of the nineteenth century, Germany developed an extensive synthetic organic chemical industry directed toward the production of pigments and dyestuffs. A high incidence of cancer of the bladder was soon noted among workers in the dye plants, which was at first attributed to aniline but later shown to be associated with organic amines. Studies at the time indicated that, while the incidence of bladder cancer was less than 1 percent for the general public, it was 25 percent for workers exposed to amines, 55 percent for those exposed to naphthylamines, and 100 percent for distillers of naphthylamines (who were potentially exposed to more of the hot vapors). These statistics brought about the first widespread recognition of the potential hazards associated with the handling of industrial chemicals.

The chemical industry in the United States did not really get under way until World War I. Until then, we were largely dependent on Germany for such things as stainless steel, dyestuffs, potash fertilizers, and many essential raw materials and finished products. Industrial safety and pollution controls in this country were virtually nonexistent; thus, at the Fourth International Congress on Occupational Accidents and Diseases held in Brussels in 1910, the entire United States activity was dismissed with the statement, "It is well known that there is no industrial hygiene in the United States." In fact, the typical American attitude at that time is perhaps best illustrated in Booth Tarkington's classic novel, *The Turmoil* (Harper & Brothers, New York, 1914), which pictures the growth of an industrial town in Indiana. When the town's leading industrialist is faced with a committee of complaining housewives, he replies:

> Smoke's what brings your husbands' money home on Saturday night. . . . Smoke means good health: it makes people wash more. . . . You go home and ask your husbands what smoke puts in their pockets out o' the pay-roll—and you'll come around next time to get me to turn out more smoke instead o' chokin' it off!

By the second decade of the twentieth century, however, gradual improvements were being made in working conditions in this country. Phosphorus poisoning in the match industry was eliminated; lead poisoning among pottery workers and painters was recognized, and crude safety precautions were introduced. New York City's spectacular Triangle Waist Factory fire in 1911 killed

145 workers, mostly young women, and vividly demonstrated the deplorable conditions in many factories. State laws were passed prohibiting employment of children under 14 in factories, the employment of women was limited to an 8-hour day, and the first workmen's compensation laws were introduced. When Franklin Roosevelt and his New Deal swept into power in 1932, the 40-hour week became standard, and stricter regulations plus stronger labor unions brought about substantial improvements in working conditions as well as a widespread consciousness of the need to protect the health of employees from potential hazards in every occupation.

Pollution Control

In a discussion of chemical pollution, we must remember that *any* material (even water or table salt) can be a poison if taken in too large quantities. There is always some uncertainty in establishing hazardous limits and, in some cases, even some question as to whether or not a given material represents any hazard at all. Chemical pollutants can be divided roughly into two classes, *degradable* and *nondegradable* pollutants. The degradable pollutants are those that are either part of the natural environment or are easily broken down to components of the environment. Nondegradable pollutants are those that are not part of the natural environment and are not easily broken down when released to the environment. Obviously, there can be no sharp dividing line.

Enforcement of pollution standards in the United States now comes under the control of three federal government agencies. The Occupational Safety and Health Administration (OSHA) was established in 1970 to develop occupational safety and health standards, to conduct inspections, to determine the status of compliance with their standards, and to issue citations for noncompliance. It is probably safe to say that most government agencies are not popular these days—it is also safe to say that with most industrialists, OSHA rates at the bottom of the list. While OSHA is presumably interpreting the wishes of Congress, which bears ultimate responsibility for our bureaucratic system, there seems to be no question that OSHA has frequently antagonized industry by unannounced inspections and often arbitrary establishment of standards which have proven very expensive and difficult to meet.

The second government agency concerned with pollution control is the Environmental Protection Agency (EPA). Also established in 1970, EPA has the mission to abate and control pollution, establish standards, and enforce pollution regulations. Its broad spectrum of activity includes standards for air and water pollution, automobile exhausts, noise abatement, ocean dumping, radiation protection, pesticides, and the disposal of solid and chemical wastes.

The mining industry comes under the control of the Mine Safety and Health Administration (MSHA), established in 1977 with the mission to develop mandatory safety and health standards, ensure compliance with such standards, and prevent and reduce mine accidents and occupational diseases in the industry.

Occupational Pollution

The field of pollution control is extremely broad and involves thousands of regulations covering many different materials. We can only discuss a few of the most significant aspects now affecting industry in the United States. Occupational pollution has been with us for over 2000 years, and it is not a story that industry can be proud of! Only too often have profits been placed ahead of safe working conditions. Industrial hazards have been downplayed and, in some instances, information about potential hazards has been deliberately withheld from employees. Again, it is simply "the tragedy of the commons." To provide safe working conditions, a manufacturer must spend money from his own pocket; in the past it was more profitable not to spend this money since the cost of on-the-job injuries was simply dumped on "society," thereby relieving the manufacturer of any responsibility.

One of our most hazardous occupations continues to be underground coal mining, chiefly because the inhalation of coal dust over a period of time causes *black lung disease* (pneumoconiosis). There are presently over 100,000 identified cases of black lung disease in the United States, and there are some 4000 deaths per year in which this is a chief or contributing cause. Billions of dollars are now being spent annually to compensate victims of black lung disease and their dependents. Much of this human misery and expense could be prevented by maintaining strict dust-level standards for mines. There is also no question that the other hazards of coal mining could be greatly reduced; studies have shown that the accident rate in the safest mines is no greater than that in other occupations (such as retail trade), which are not normally considered to be dangerous.

A similar dust disease, *brown lung* (byssinosis), affects workers with cotton, flax, and hemp. It is estimated that as many as 17,000 cases of brown lung disease now exist in the United States. There is some uncertainty in this figure because some manufacturers have been reluctant to even admit the existence of the disease and have strongly opposed government attempts to regulate working conditions or even to evaluate the health hazards of this industry.

A third dust disease, recently found to be extremely serious, affects workers with *asbestos.* Asbestosis, a gradual loss of lung capacity, has been known for many years; as early as 1918, insurance companies were turning down asbestos workers because of the hazardous nature of their work. It has now been found, however, that exposure to asbestos also causes cancer of the lungs and of the membranes that line the chest and abdominal cavities. The hazard is particularly acute for those who are also smokers. Since asbestos has been widely used for water pipes, brake linings, insulation, and wall and ceiling coatings (in view of its fire-resistant quality), it is estimated that some half million workers have received excessive exposure to asbestos dust—which may eventually result in 100,000 deaths from lung cancer and 70,000 deaths from other asbestos-related diseases. There is, of course, no way to estimate the numbers of children and adults who are daily exposed to asbestos dust in old buildings of all kinds, including school buildings and theaters, but public awareness of the problem is helping to eliminate much of this risk.

In the chemical industry itself, standards have been set by OSHA for maximum permissible exposure to all the common chemicals that are produced commercially. In addition, there are special regulations regarding exposure to inorganic chemicals such as lead and arsenic and for exposure to carcinogens such as vinyl chloride and acrylonitrile.

Two illustrate the difficulty of establishing safe levels of exposure, consider the case of *vinyl chloride,* which has been produced in very large quantities in Germany and the United States since the early 1930s. Acute animal toxicity to vinyl chloride was demonstrated as early as 1938; in 1949, liver damage was discovered in fifteen of forty-eight workers in the Soviet Union. Despite these facts, industrial exposure limits for many years were set at 500 ppm. In 1974, Dr. John Creech, Jr., a physician for the B. F. Goodrich chemical plant in Louisville, Kentucky, found a large number of serious liver ailments, including several cases of a very rare liver cancer, among workers exposed to vinyl chloride. An investigation of other manufacturers of vinyl chloride revealed that similar conditions existed throughout the industry. Federal standards for exposure to vinyl chloride were reduced from 500 to 1 ppm, and additional safety regulations have been instituted for all manufacturers of vinyl chloride.

In 1977, an interesting situation developed with respect to industrial exposure to *benzene.* A report from the Department of Health, Education, and Welfare appeared, alleging that workers exposed to benzene had about five times the probability of dying from leukemia (a form of cancer) as the rest of the population. Seizing upon this, OSHA lowered the limits for benzene exposure from 10 to 1 ppm. This was an extremely serious matter to the chemical industry, because benzene is widely used as a solvent and as a raw material for the production of petrochemicals and polymers. Because of its ability to raise octane number, substantial quantities of benzene are present in gasoline; human exposure to gasoline fumes automatically means exposure to benzene vapors. Benzene is also widely used in industrial and university research laboratories—in fact, for many years practically every university chemical engineering department used a mixture of benzene and toluene for its undergraduate laboratory experiment in distillation.

This arbitrary decision by OSHA was taken to court by the chemical industry, which pointed out that the reduced benzene standard would necessitate an immediate expenditure of $500 million and would probably force some of the manufacturers out of business. OSHA admitted that there was no evidence to show how many lives would be saved or whether, in fact, the existing standard of 10 ppm was measurably unsafe. They claimed, however, that they were empowered to set standards regardless of economic impact. The case was taken all the way to the United States Supreme Court, with the position of the chemical industry being upheld at all levels. OSHA had exceeded its authority because it was unable to show that the 1 ppm limit was "reasonably necessary or appropriate to provide safe and healthful employment." The point to remember is that OSHA does not have the authority to create absolutely risk-free working conditions regardless of cost.

ENVIRONMENTAL PROTECTION

All matters of environmental pollution fall under the control of the Environmental Protection Agency. Their activities are very broad and include the following general categories: (1) control of air pollution and establishment of national air standards; (2) control of water pollution, including chemical wastes, radioactive wastes, and thermal pollution; (3) garbage disposal and the management of land-fill areas; (4) control of pesticides and enforcement of the Federal Insecticide, Fungicide, and Rodenticide Act; (5) building and highway construction; (6) protection of wetlands; (7) noise emission standards for trucks, railroads, and construction equipment; (8) ocean dumping; and (9) discharge of oil and hydrocarbons to the environment.

Since the establishment of EPA, an *environmental impact study* is required before anyone can do anything that materially changes the natural environment. In some cases, the studies themselves are very detailed and costly. There is no question that environmental regulations have become a very serious and, in some instances, an overriding consideration in the location of industrial plants today and sometimes even in the selection of the particular manufacturing processes that are used. Environmental considerations have led to the closing of many plants, have greatly increased the cost of many other operations, and have inhibited the growth of much of the chemical, petroleum, and mining industries in recent years.

Air Pollution

Carbon monoxide gas is the air pollutant produced in the largest quantities by human activity and on a global scale exceeds the mass of all other air pollutants combined. Total worldwide production of carbon monoxide is roughly 400 million tons per year, with some 55 percent coming from motor vehicle emissions. The carbon monoxide content of our atmosphere is quite variable, of course, ranging from about 1 to 70 ppm in built-up areas, with brief peak levels as high as 140 ppm. Concentrations over 600 ppm may cause headaches and nausea, and prolonged exposure to concentrations over 1000 ppm may be fatal.

In spite of the large amount of carbon monoxide introduced into the atmosphere by human activity, it is now believed that natural sources of carbon monoxide contribute roughly ten times as much. It is produced by electrical storms, by forest and prairie fires, and by the decomposition of plant and animal materials. Methane and other hydrocarbons released to the atmosphere can react with oxygen to form carbon monoxide. Large quantities are also produced in the oceans by marine organisms such as jellyfish and by the decomposition of organic material in surface waters.

Even with this tremendous and continuous pollution of the atmosphere by carbon monoxide, the concentration in "clean" air remains extremely low (0.2 ppm or less) and has been unchanged for a century or more. Obviously there are natural processes that rapidly and effectively remove carbon monoxide from the atmosphere. These are not completely understood, but it is known that sunlight can activate the oxidation of carbon monoxide to carbon dioxide, particularly in the upper atmosphere or *stratosphere.* There are also friendly soil bacteria and fungi that effectively remove carbon monoxide from the air, and it is now believed that the capacity of the soils in the United States is sufficient to take care of our entire production of carbon monoxide. Without these natural removal processes, the natural production of carbon monoxide alone would be enough to build up a lethal concentration in the atmosphere in a period of some 2000 years.

The physiological effects of carbon monoxide result from its strong affinity for hemoglobin, roughly 300 times that of oxygen. Exposure to carbon monoxide thus gradually cuts off oxygen transfer to the body and can cause death by asphyxiation. The process is entirely reversible, however; carbon monoxide is not a cumulative poison and is easily removed from the blood when the individual is exposed to pure air or oxygen. Many individuals, such as traffic officers in vehicular tunnels, have been exposed to high concentrations of carbon monoxide for much of their lives without known harm. Except possibly for those who live in very high traffic areas, then, carbon monoxide cannot be considered a serious environmental pollutant.

Sulfur dioxide is quite another story. Annual worldwide emissions of sulfur dioxide by human activities now total about 175 million tons, with some 70 percent coming from the burning of coal and 16 percent from the burning of petroleum products, particularly the high-sulfur residual fuel oil used by many industrial plants. Sulfur dioxide emissions have grown steadily for the past thirty years but are now expected to level off and perhaps even decrease somewhat as greater restrictions are placed on industrial pollution.

Sulfur dioxide is recognized as an irritant gas, causing problems for those with asthma, emphysema, or other respiratory diseases. In high concentrations, it is very harmful to vegetation. This has been demonstrated near mining towns where sulfide ores were roasted to burn off the sulfur and the sulfur dioxide gases simply discharged to the stack; Sudbury, Ontario, and certain mining towns of the western United States are surrounded by large and desolate areas filled with dead and dying trees and other vegetation.

Sulfur dioxide is removed rapidly from the atmosphere. Quite soluble in water, it forms a dilute solution of sulfurous acid and is thus readily removed in rainfall as acid rain. If oxides of nitrogen are present in the atmosphere, sulfur dioxide is easily oxidized to form sulfuric acid as well, thus increasing the incidence of acid rain. The problems brought about by acid rain can originate many hundreds of miles away, making the control of this form of pollution extremely difficult. Structural materials, particularly limestone, may be corroded; however, the increased acidity of water and soil represent the most serious aspect of the problem, affecting the growth of crops and killing fish. Largely because of sulfur dioxide emissions by our midwestern utilities, acid rain is particularly bad in the northeastern United States and in southeastern Canada. For example, the Great Lakes states of Illinois, Indiana, Michigan, Ohio, and Wisconsin burned roughly 37 million tons of coal to produce electricity in 1949. By 1978, this had increased to 145 million tons, almost four times the postwar figure. Acid rain is not exclusively an American phenomenon; it has now been found in much of northern Europe, where countries such as Sweden and Norway are being affected by sulfur dioxide emissions that originate in the United Kingdom.

It is of interest to note that, in addition to the sulfur emitted to our atmosphere by human activities, an estimated 100 million tons of hydrogen sulfide gas (familiar to most of us as the "rotten egg" gas that also tarnishes silver) are contributed annually by natural biological decay processes occurring on land and in the ocean. It is also estimated that some 44 million tons of sulfur per year enter the atmosphere from sea spray, largely in the form of minerals such as calcium sulfate. The total amount of sulfur emitted to the atmosphere by these natural processes is actually close to the amount from human activities.

Oxides of nitrogen, which are considered to be an important source of smog, can be produced in any combustion process where the temperature is high enough to "fix" the nitrogen in the air by causing it to react directly with oxygen. The initial product is nitric oxide (NO), but on cooling this gradually oxidizes to nitrogen dioxide (NO_2). Normally there will be a mixture of the two present in exit gases, and this is sometimes referred to as NO_X. It is estimated that about 60 million tons of nitrogen oxides per year are produced by human activities, with roughly half coming from the combustion of coal.

Nitrogen dioxide reacts readily with moisture to form nitric acid, which is easily removed from the atmosphere by absorption in rainfall. Since oxides of nitrogen are normally present in only fairly small concentrations, and since they are quickly removed from the atmosphere, they seldom present a physiological hazard. They can, however, be a problem if organic chemicals are also present since they promote the formation of smog.

Nitrogen is, of course, essential for plant growth. The *nitrogen cycle* is a critical element in all biological processes, and it is estimated that the amount of nitrogen dioxide released to the atmosphere through natural biological action is about fifteen times that contributed by industrial pollution and automobile exhaust. In addition, biological reactions release even greater quantities of nitrogen in the form of ammonia and nitrous oxide (N_2O). As far as the nitrogen cycle is concerned, then, pollution sources are not very significant.

Ozone (O_3) is naturally present in the atmosphere in very small concentrations. It is produced by the action of sunlight on atmospheric oxygen, and the concentration of ozone is thus higher in sunny areas such as southern California. Although ozone can have detrimental effects on vegetation, the amount usually present in our atmosphere is not in itself physiologically harmful. Like the oxides of nitrogen, however, ozone does promote the formation of smog when organic chemicals are also present in the air.

There has been considerable concern in recent years about the amount of ozone present in the upper atmosphere or stratosphere. In the intense sunlight, the ozone concentration in the stratosphere reaches a value of about 15 ppm, some 1000 times its value at sea level. This *ozone layer* is valuable to us for two reasons. As we have seen, it absorbs practically all of the most harmful shortwave ultraviolet light present in sunlight. It has been estimated that any substantial reduction of the ozone layer could produce tens of thousands of additional cases of skin cancer, with many fatalities. In addition, since ozone is a good absorber of infrared radiation, any change in the ozone concentration could affect the heating of the entire stratosphere and bring about undesirable changes in our global environment as well.

Much of the opposition to the supersonic transport plane was based on concern that water vapor and oxides of nitrogen ejected into the stratosphere could act to destroy the ozone layer. Subsequently, it was found that chlorine gas is even more effective in destroying ozone. Elemental chlorine itself is no hazard, since it is rapidly removed from the atmosphere by absorption in water or by reaction with ammonia vapor. It was pointed out, however, that millions of tons of chlorine-containing fluorocarbons had been released to the atmosphere through the use of push-button sprays; since these compounds are chemically inert, they can gradually migrate to the stratosphere where the intense sunlight can dissociate them into free chlorine.

A Federal Task Force on Inadvertent Modification of the Stratosphere issued a report in 1975 which ". . . concluded that fluorocarbons released to the environment are a legitimate cause for concern.. . . It would seem necessary to restrict uses of fluorocarbons-11 and -12 to closed recycled systems or to other uses not involving release to the atmosphere." Although there is a legitimate basis for questioning these findings, we can very easily get along without using fluorocarbons in push-button sprays, and they obviously should be done away with if there is even the remotest possibility of harm to the ozone layer. Industry has finally yielded to the pressure and discontinued the use of fluorocarbons for aerosols and other applications that involve release to the atmosphere.

Large quantities of *hydrocarbons* are emitted naturally into the atmosphere from forests, vegetation and (in the form of methane) from bacterial decomposition of dead organic matter. Human activity accounts for only some 15 to 20 percent of the total quantity of hydrocarbons entering the atmosphere, but this is concentrated in urban areas. The chief source of hydrocarbon pollution is automobile exhaust resulting from incomplete combustion of the fuel; other sources include power plants, chemical plants, petroleum refineries, and the burning of solid wastes. It has been estimated, for example, that as much as 2 percent of our total production of gasoline is lost by volatization during transfer processes of one kind or another. The evaporation of dry-cleaning solvents is estimated to be at least 100,000 tons per year. There are also tremendous amounts of organic solvents released to the atmosphere from paints, lacquers, adhesives, printing, and push-button sprays.

Hydrocarbons are rapidly oxidized in the atmosphere to carbon monoxide and carbon dioxide, which can then be removed by the usual natural processes. In addition, there is evidence that some types of soil bacteria rapidly consume methane and probably other hydrocarbons that may be present in the air. The most critical hydrocarbons are the "reactive" ones that can be converted to smog and those that are either themselves carcinogens or that can be converted to carcinogens. It is fortunate that methane, our most common hydrocarbon pollutant, falls into neither of these categories.

Under normal conditions, large cities are able to get rid of their hydrocarbons and other airborne pollutants merely by discharging them to the atmosphere and allowing them to go wherever the wind blows. This produces plumes of airborne dirt and smoke which extend as visible streamers for many miles downwind of their source. Plumes originating in the metropolitan New York City–northeastern New Jersey complex, for example, may be found in the upper Hudson Valley, in southeastern New England, and over the Atlantic Ocean. Commercial airline pilots flying the Atlantic are often able to spot these plumes hundreds of miles at sea. Interestingly, Connecticut has come under pressure from the Environmental Protection Agency to clean up its air—even though it has been shown that stopping all industry and all automobile traffic in that state would not solve the problem because of the air pollution coming in from the New York City area.

The word *smog* was originally coined to describe a mixture of smoke and fog. Today it is used more often to categorize the *photochemical* smog produced by reactions among hydrocarbons, oxides of nitrogen, and ozone that are catalyzed by ultraviolet light from the sun. Smog is a particular problem for cities such as Los Angeles, Denver, and Salt Lake City, where there is much sunlight and where there are geographic barriers to prevent the normal movement of air. Smog causes severe eye irritation and difficulty in breathing; it is a hazard to the very young, the elderly, and to anyone affected by respiratory problems.

The most serious consequences of smog have occurred during periods of *temperature inversion.* Normally, the temperature of the atmosphere decreases with altitude; since hot air rises, the discharge from a chimney simply continues upward until it is gradually dissipated in the atmosphere. A temperature inversion occurs when warm air "overruns" cold air, so that the temperature increases with altitude. The discharged pollutants are now trapped below the inversion level with no place to go, and they can rapidly build up to unpleasant or even dangerous levels.

There have been several well-publicized instances when temperature inversions have had serious consequences. The factory town of Donora, Pennsylvania is located in a steep valley of the Monongahela River; when a smog persisted for five days in 1948, roughly half of its inhabitants became ill, and there were twenty deaths. In London, a combination of fog and coal-smoke which lasted for several days in 1952 caused such a dense smog that people could barely see their own feet (some even walked off the docks and fell into the Thames River). Mortality figures included some 4000 "excess" deaths that could be attributed to the smog. Los Angeles has periodic smog alerts, and a severe temperature inversion in November of 1966 caused New York City to issue its first dangerous smog warnings.

The problem of smog is being attacked on several fronts. Industrial pollution has been substantially reduced. The burning of garbage and other wastes has been essentially stopped. Automobile emission controls are reducing hydrocarbon emissions somewhat, and better-mileage cars and reduced gasoline consumption in general will continue to help. With a constant or possibly reduced consumption of energy in our future, we can expect the smog problem will eventually disappear.

For many years, both tetraethyl and tetramethyl *lead* have been used widely as gasoline additives to improve octane number. In the 1970s, automobile emissions of lead to the atmosphere reached about half a million tons annually, corresponding to over 10 percent of the entire world production of lead. Most of this lead entering the atmosphere has been deposited in our oceans, either directly or through runoff to streams and rivers. Recent experiments have shown the lead concentration in ocean surface waters of the Northern Hemisphere to be several times that in unpolluted waters of the Southern Hemisphere.

If taken in sufficient quantities, lead is a known biological poison. There is even some evidence that the use of lead utensils for cooking and eating may have been a factor in the downfall of the Greeks and Romans. Since the lead emitted by automobile exhausts is normally in a volatile form, much of it tends to remain in the atmosphere around cities and highways, where atmospheric levels often exceed EPA standards. All of us ingest a certain amount of lead in our food and drinking water, but there is no really good evidence to show that atmospheric levels of lead have been harmful thus far, even in large cities with heavy automobile traffic.

As a result of the introduction of automobiles that require lead-free gasoline, of course, the problem of atmospheric lead is beginning to lessen. Since lead in the oceans is slowly removed by biological processes, worldwide levels of lead in the environment should begin to drop slowly over the next ten or twenty years.

Water Pollution

Control of water pollution can be achieved through careful monitoring and treatment of waste streams. Some impurities can be destroyed chemically, others removed by filtration. Heavy metals can often be removed by ion-exchange processes similar to those used to "soften" water. Organic materials can sometimes be separated out by settling and the recovered organics can often even be burned as fuel. The point is that any chemical waste stream can be cleaned up—it is simply a matter of price.

On a global scale, one of the most serious water pollutants is *mercury,* which seems to have increased sharply in the past twenty to thirty years. Measurements of the mercury content of snow on the Greenland glacier, for example, have shown more than a doubling. Both elemental mercury and several of its compounds are fairly volatile and pass readily into the atmosphere. The natural flow of mercury into the air is estimated at about 100,000 tons a year, chiefly through the "degassing" of the earth's crust. Evidently human activities such as agriculture, mining, and construction keep exposing more and more of this material, thereby favoring the release of mercury vapor and volatile mercury compounds to the atmosphere.

Small traces of mercury are present in coal and petroleum, and most of this mercury enters the atmosphere when they are burned. Many mineral ores also contain substantial quantities of mercury, which is volatilized when the ores are roasted. These sources, however, are estimated to be no more than one-tenth of the natural emissions.

Among the chief industrial uses of mercury is the manufacture of chlorine and caustic soda by the electrolysis of brine. Organic mercury compounds are also used as an agricultural fungicide in the treatment of seeds, and deaths have been recorded from eating treated grains. Environmental problems with mercury are the direct result of irresponsible discharge of mercury-containing effluents by chemical companies. The worst example was in the 1950s in Minimata, Japan, where the Chisso Chemical Corporation discharged large quantities of volatile methyl mercury into the environment. Thousands of people were affected, and there were at least forty-six recorded deaths.

Another problem with the discharge of even low-level mercury wastes is caused by the fact that microorganisms in river and ocean sediments can convert inorganic mercury into volatile organic mercury compounds. Furthermore, these organic mercury compounds become concentrated as they move up the food chain, resulting in high concentrations of mercury in some types of fish. In Sweden, it has been necessary to stop fishing in certain areas because of mercury contamination from agricultural runoff. In the United States, it was found during the 1970s that contamination from chemical plants was causing concentrations of mercury in tuna and swordfish above the Food and Drug Administration limit of 0.5 ppm.

Other metals that are potential hazards to water include cadmium, nickel, zinc, chromium, and arsenic. The most serious of these is cadmium, a cumulative poison which is widely used in electroplating and for the manufacture of pigments, alloys, and batteries. Cadmium poisoning has occurred in industrial workers in Japan and in villages there where the water supply has been contaminated by drainage from cadmium mines. Even small amounts in ocean water can present problems, because cadmium is strongly concentrated in marine life such as mollusks. It has not been a problem in the United States.

Sometimes the chemical industry has pollution problems that present a real challenge. A particularly interesting example is in the Solvay process for the manufacture of soda ash. About a ton of calcium chloride waste is produced for every ton of product, and calcium chloride has no commercial value; millions of tons have been dumped on the open land or discharged to rivers or quarries where it has contaminated drinking water with a saline taste and killed vegetation. Enforcement of environmental restrictions has now closed down almost the entire industry, and we will have to rely on natural soda ash from the Wyoming deposits for our future requirements.

Another specialized problem of water pollution is the case of mine drainage, particularly from underground coal mines. Eastern coal has appreciable quantities of iron sulfide (pyrites) which is easily oxidized by air to form sulfuric acid. This dissolves in ground waters, polluting streams and killing fish and vegetation. Mine drainage is a particular problem in Appalachia, where many small communities are located on the banks of polluted streams.

NONDEGRADABLE POLLUTANTS

An extremely serious pollution problem today is disposal of the millions of tons of nondegradable synthetic organic polymers that we are producing annually. Unlike the natural polymers, since there are no microorganisms that will consume them, materials such as nylon, polyethylene, polystyrene, and others in common use persist for very long periods of time once they are placed in the environment. It is expected that they may slowly degrade and crumble over a number of centuries, but this is not a satisfactory answer in view of the tremendous quantities involved.

One possibility that has been considered would be to convert them back to simple compounds by burning. With the materials very widely distributed and in many different forms, collecting them would be very difficult; in addition, as noted earlier, they often produce effluent gases that are toxic and/or carcinogenic. Burning, then, is not a promising solution, and our best choice here is to minimize our dependence on these nondegradable synthetic materials.

Although the synthetic polymers produced today are nondegradable, it is possible to produce synthetic organic polymers that are degradable by modifying the arrangement of the atoms in the polymer chains. Considerable research is now being directed toward the production of satisfactory degradable polymers. If they can be produced commercially, they could begin to replace some of today's nondegradable polymers; certainly, as a minimum, such degradable polymers should be required for all packaging materials.

To illustrate the unexpected problems that can be encountered in the use of synthetic organic chemicals, consider the case of the PCBs (polychlorinated biphenyls). These highly chlorinated synthetic chemicals were first produced in 1929 and proved effective as heat-transfer fluids, high-temperature lubricants, and in the manufacture of paints, varnishes, adhesives, and waxes. Because they were nonflammable and had good electrical properties, they were widely used as heat-transfer fluids for large electrical transformers and capacitors. They were not considered to be particularly hazardous at first; large amounts escaped to the environment from open burning and incineration of wastes, vaporization of paints, coatings and plastics, and from simply dumping PCB wastes into rivers and lakes.

In the early 1960s, evidence began to accumulate to indicate that PCBs were becoming a hazard. They were extremely persistent chemicals when discharged to the environment and were being distributed very rapidly throughout the entire Northern Hemisphere. When ingested, they concentrated in fat, and experiments with animals showed them to be carcinogenic. High levels of PCBs were soon found in fish and in other organisms collected in the open sea, miles from land. Traces began to appear in human fat and milk. In 1968, more than 1000 people in Japan became violently ill when exposed to several ppm of PCBs from a heat exchanger that had leaked into the oil used for cooking food.

Because of adverse environmental effects, United States production of PCBs was voluntarily terminated in the 1970s. With the passage of the Toxic Substances Control Act of 1976, the manufacture and use of PCBs in the United States has been essentially outlawed. Unfortunately, there are still areas badly contaminated with PCBs, and it will be a long time before they disappear entirely from our environment.

The Kepone Story

The story of Kepone represents one of the worst instances of pollution that has ever occurred in the United States. Kepone was a highly chlorinated synthetic organic pesticide developed by Allied Chemical and found to be very effective in killing fire ants and household roaches. (As it later developed, tests had also shown quite clearly then that it was toxic to humans.) After considerable developmental work, Allied Chemical arranged for two of their employees to form the Life Science Products Company, a "spinoff" firm to manufacture Kepone. A former gasoline service station located a short distance from the main Allied Chemical plant at Hopewell, Virginia, was purchased, and production of Kepone was begun there, with Allied Chemical furnishing the raw materials and helping with the marketing of the final product.

In June of 1975, Dale Gilbert, a 34-year-old employee of Life Science went to the company doctor complaining of tremors and pains. He was told this was probably due to stress and given a prescription for a tranquilizer. Dissatisfied with the diagnosis, Gilbert's wife set up an appointment with a local cardiologist, Dr. I-Nan Chou. When Dr. Chou sent a blood sample to the Center for Disease Control in Atlanta, he was told that it contained a high concentration of Kepone.

Shortly after this, Dr. Robert S. Jackson, a state epidemiologist, paid a visit to the Life Science plant, where he found employees literally immersed in the white powdery chemical. "Kepone was everywhere. They were sloshing around in it with no boots, gloves, or respirators on." He found seven workers so affected that he hospitalized them on the spot. Later tests showed the presence of Kepone in the blood of almost 100 persons, including employees, former employees, wives, and children. Of five known pregnancies among Life Science personnel, two resulted in stillbirths and one in spontaneous abortion. One 4-month-old child was found to have traces of Kepone in his blood apparently because his clothing was laundered in the same washing machine as that of his father.

Shortly after the start of Kepone production, digesters at the local sewage treatment plant had failed. The Kepone wastes being discharged to the sewers had killed the bacteria necessary to oxidize the sewage, and the town was left with large quantities of Kepone-contaminated sludge. An estimated 100,000 pounds of Kepone had also been dumped into the James River and for many months, fishing in the James and in parts of Chesapeake Bay was forbidden because of Kepone contamination. In May of 1976, a federal grand jury indicted Allied Chemical, alleging 1094 separate criminal charges in the discharging of Kepone into the James River. Pleading *nolo contendere,* Allied was fined over $13 million, the maximum allowed under the statutes. Life Science employees and fishermen deprived of their fishing grounds brought civil suits against Allied for some $160 million. The judge handling the cases strongly advised uncontested settlement of claims and full cooperation in any steps necessary to clean up the environment.

The Love Canal Tragedy

About 100 years ago, William Love conceived the idea of a "model city" program for Niagara Falls, New York. He was granted a franchise by the state in 1891 to dig a 6-mile canal from the Niagara River to provide water and power, with discharge to be at the "Devil's Hole," a high point on the bank of the Niagara Gorge midway between the city of Niagara Falls and the village of Lewiston. Ground was broken, construction of the canal began, and "a good farm was turned upside down." The project was never completed, however, and some years later the property was acquired by Hooker Chemical Company, which had large chemical manufacturing plants based on the cheap hydroelectricity from Niagara Falls.

At some time in the 1940s, Hooker started to use the old canal bed as a dump for toxic wastes, and thousands of drums were dropped directly into the receding water or buried in its banks. In 1953, Hooker sold the land to the city's Board of Education for $1. The canal was filled with dirt, a school was built, and lots were later sold by developers. Houses were built along the banks of the old canal with their back yards directly over the area of the former canal bed.

In 1976, after six years of abnormally heavy rains, the old canal "overflowed its underground banks," and chemicals started surfacing. From time to time children and dogs playing over the former excavation received chemical burns, and it has also been alleged that a number of miscarriages and birth defects resulted among families living near the canal. One of the property owners was Karen Schroeder, the mother of four children, of whom one had been born with a cleft palate, an extra row of teeth, and slight retardation. In the summer of 1978, public concern about the safety of the neighborhood was aroused when Mrs. Schroeder's swimming pool was literally popped out of the ground by the rising water; her garden was killed, and the redwood posts of her backyard fence were eaten away by oozing chemicals. Local authorities pumped from her yard some 17,500 gallons of water so badly polluted that the local waste disposal companies refused to handle it.

Tests of the water and air above Love Canal revealed the presence of eighty-two different chemical compounds, including eleven suspected carcinogens. Furthermore, material leached from the buried drums was leaking through the basement walls of some 100 of the homes and polluting the air inside the houses. New York State Health Commissioner Robert Whalen declared a health emergency, and pregnant women and infants were advised to leave the area immediately. President Carter declared a federal emergency and provided funds for emergency resettling of the residents. Eventually 237 families were evacuated from homes bordering the canal; the houses were purchased by the state, the area was surrounded by a fence, and bulldozers were brought in to clean up the mess. The total cost was about $23 million. In December, 1979, the Justice Department filed a $124.5 million suit against Hooker Chemical for dumping toxic wastes at four sites in Niagara Falls, including Love Canal. In addition, some $25 million in damage suits has been filed against New York State by former residents of Love Canal.

In January 1981, the Justice Department negotiated an agreement whereby Hooker Chemical agreed to spend $15 million to clean up and monitor its chemical dump site in the Hyde Park area of Niagara Falls. Hooker admitted to depositing 80,000 tons of chemical wastes at Hyde Park between 1953 and 1974, of which 80 percent were toxic. This settles only one of four suits filed by the Justice Department against Hooker charging violations of federal laws in the disposal of chemical wastes at the four sites in the Niagara Falls area, including Love Canal.

As in the case of Three Mile Island, by far the most serious consequences of Love Canal have been the damage to the residents' peace of mind and to confidence in government agencies. A committee of scientists under the chairmanship of Dr. Lewis Thomas of the Sloan-Kettering Cancer Center found that bungling by public and private investigators ". . . fueled rather than resolved public anxiety." Much of the original publicity about Love Canal was specifically designed to obtain federal disaster funds; by inflaming the community, it made objective scientific studies very difficult. In many instances, too, state and federal officials were acting independently of each other. One study of alleged chromosome damage was so poorly designed that it never should have been undertaken. After two years of intense study, there is still no clear evidence to indicate whether or not any of the residents of Love Canal were actually harmed by the chemical wastes. In view, however, of the psychological damage they have suffered and will carry for the rest of their lives, these people are certainly entitled to the special help they have received.

Love Canal did have one positive consequence—it dramatically illustrated the potential hazard of chemical waste disposal. It has undoubtedly been the stimulus for the decision of Congress to establish a $1.6 billion fund for the cleanup of chemical waste dumps, with 88 percent of the cost coming from industry. It is hoped the lesson of Love Canal has been good education for the chemical industry, the government regulatory agencies, the public, and the news media.

Radioactive Wastes

The problem of high-level radioactive wastes has been discussed in Chapter 4 and will not be repeated here. Low-level radioactive wastes are those associated with contaminated clothing, liquid and gaseous effluents from nuclear power plants, radioactive isotopes (such as iodine 131 used by hospitals), radioactive tracers used by university and industrial research laboratories, and the like. In general, these do not present a serious disposal problem. Liquids and gases are commonly diluted and discharged to the environment, provided the radioactivity is sufficiently low; solid wastes are normally compacted and buried.

It is ironic to note that several states, such as Connecticut, which are clearly reaping the benefits of nuclear power, have themselves taken a very arbitrary position with respect to radioactive wastes—"Not in my backyard, you don't!" The few states that do accept radioactive wastes today are understandably less than enthusiastic about becoming dumping grounds for the nation, and it is possible that the federal government may eventually be forced to take over the disposal of all radioactive wastes.

The inevitable leveling-off of material growth in the United States will in itself help to end the growth of most forms of pollution. It is impossible to carry out any industrial operation without some risk and some pollution, and we must remember that establishment of optimal pollution levels—reasonable and acceptable from an environmental, social, economic, and technical point of view—will involve choices, perhaps none of them entirely satisfactory. Those who demand zero risk and zero pollution are saying, in effect, that we should give up our modern technology and return to living in caves.

Chapter Seven

The Bottom Line

Although money itself is not a resource, it represents the value placed on our resources and serves as a measure of their availability to us. The cost of everything we have depends on the cost of obtaining basic raw materials and the costs involved in converting them to finished goods. It is impossible to talk about our future demands for resources without also considering the cost of energy, the cost of labor, and the cost of transportation.

The situation we are facing today is clearly illustrated in Figure 25, which gives United States *disposable income* per capita adjusted for inflation and covering the period from 1960 to 1980. Disposable income is simply the per capita average of personal income less personal tax payments. It is the income per capita available for spending or for saving and is probably the best measure of our standard of living.*

Between 1960 and 1973, disposable income per capita increased more or less regularly at a rate of 3 percent per year. In other words, the average American enjoyed a 50 percent rise in his or her standard of living during this period. When the Arab oil embargo and resulting increase in oil prices brought a drop in disposable income for 1974, it took three years before we saw a return to

*Do not confuse disposable income with *discretionary* income, a term commonly used by the news media. Discretionary income is personal income less taxes and payments for necessities such as food, housing, and medical care.

"Hi! I'm Big Brother, and I'm running for President in '84."

Figure 24. (Drawing by Charles Addams, © 1980 The New Yorker Magazine, Inc.)

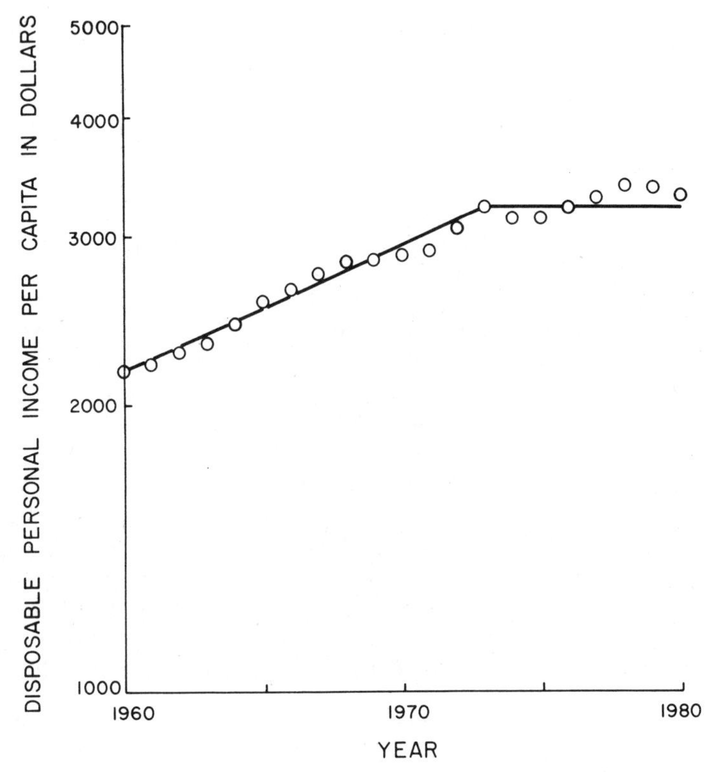

Figure 25. Disposable personal income adjusted for inflation.

its 1973 value. The years 1977 and 1978 brought small increases, but higher inflation in 1979 and 1980 led to further drops—to a value only slightly above 1973. It is possible, perhaps even probable, that the year 1973 marked the time of greatest wealth for the United States. In view of the realities in our world today (still higher oil prices, continuing inflation, unrestrained population growth), it would certainly appear wise to proceed on the assumption that we have, in fact, reached an economic plateau.

A RESOURCE-LIMITED ECONOMY

The rapid growth we experienced following World War II was spread over all of our activities. Most obvious was the rapid material growth; however, we have also enjoyed rapid growth in our educational system, in the arts, in government, in sports, in leisure time, in communication, in human knowledge, and, unfortunately, in pollution. All of these must be taken into consideration as we begin to recognize our transformation to a stationary resource-limited economy.

Think for a moment about this definition of an ideal stationary society by economist Kenneth Boulding in *The Meaning of the Twentieth Century:*

> . . . everything as it is consumed or disappears from a society is simply replaced; the whole activity of the society is devoted to replacement. The population is constant, each age group as it passes into the next or as it dies out is replaced by another. The educational process is only just sufficient to replace the loss of knowledge by death and old age. As it wears out, the physical capital of the society is simply replaced by identical objects, the production of everything is exactly equal to its consumption, and hence there is no accumulation and no change.

Although it is unlikely that any society has fulfilled all of these requirements, many Paleolithic and primitive cultures have existed for very long periods of time without substantial change. There have even been some fairly advanced societies that have existed for many centuries in a stable state. A particularly interesting example in the United States is the old-order Amish, who have deliberately rejected all modern conveniences and who practice an agricultural system that has been essentially unchanged over the past 200 or 300 years.

THE KEYNOTE CAN BE QUALITY

One of the more immediate challenges the United States faces is how to deal emotionally with the end of the material growth curve we have experienced since the first white people settled here more than 300 years ago. All forms of growth and improvement, of course, will not grind to a halt, although almost everyone is now agreed that population growth (for example) must end as soon as possible if we are to maintain today's standard of living—and if we are to overcome the urgent problems of pollution and rapid depletion of our natural resources. We have seen that it is very unlikely there will be any substantial future growth in our consumption of energy even though we can expect some of our fuels (such as coal) to grow at the expense of others (such as petroleum and natural gas). We have also probably reached a limit in our consumption of many raw materials, especially the more critical and/or energy-intensive items such as chromium, cobalt, aluminum, chlorine, and many of the petrochemicals and polymers based on petroleum and natural gas.

The leveling-off of our growth curve does not necessarily imply that we will have fewer material things or that our children will have "nothing to look forward to." If we continue to reduce waste and if we improve efficiency—for example in our use of energy—we can actually turn out more finished goods for the same consumption of energy and raw materials. If we can double the mileage of the automobiles on our highways, the same amount of gasoline will carry us twice as far. American industry has in the past few years been able to reduce the amount of energy used in manufacturing operations to a notable degree, and it is almost certain that further improvements can and will be made as the cost of energy continues to rise.

Certain products may gradually disappear from the marketplace and be replaced by others that require smaller amounts of energy and less critical raw materials. As an example, while the recycling of aluminum cans will do much to relieve the need to produce them from imported bauxite (a process that requires large amounts of electricity to reduce the aluminum oxide to aluminum metal), glass bottles made from abundant raw materials require less energy and can be easily reused. The glass container industry can thus look forward to a much better future than the aluminum industry.

As energy and raw materials become increasingly expensive, our "throwaway" economy will gradually come to an end. Consumers are already demanding more durable products: shoes that can be resoled, automobiles that are easy to maintain, classic styles of clothing that will not go out of style, appliances that are ruggedly built and easily repaired. If we could double the average life of our products, we could maintain the standard of living we enjoy today with only half the consumption of many of our critical resources.

WHAT ABOUT INFLATION?

Probably the most poorly understood and poorly handled economic problem facing the United States at this time is the growing rate of inflation—that is, the expansion of our money without a corresponding expansion of goods and services. This is illustrated in Figure 26, which shows the *consumer price index* (CPI) for the past twenty years. The data are very significant. From 1960 to 1966, the consumer price index increased at slightly over 1 percent a year, a very reasonable amount. In 1965 President Johnson decided to step up American participation in the Vietnam War, and by the time of our withdrawal in 1973, Vietnam had cost us some $139 billion. Our increased expenditures for Vietnam had the expected effect on the consumer price index: in the late 1960s, it began increasing at a faster rate, and it reached 7 percent a year for the 1974–78 period. Since 1978 it has been growing at an annual rate of over 12 percent.

What exactly does the CPI measure and how much significance can be

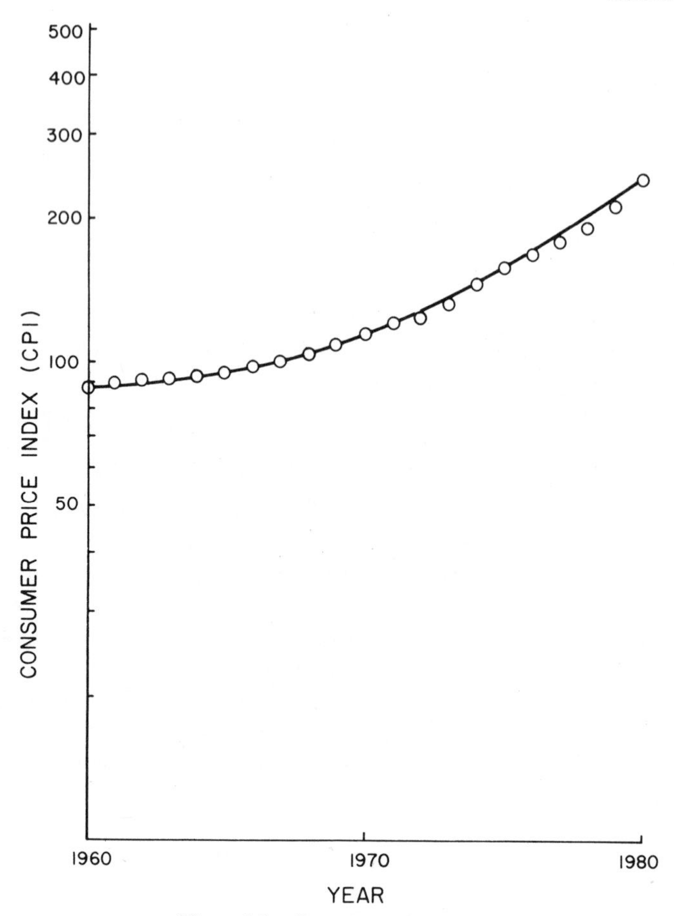

Figure 26. Consumer price index.

attached to its numbers? The cost of various items in our daily life such as food, housing, electricity, gasoline, and clothing are measured with certain weighting factors applied to determine an overall price index. Since we all have different tastes and hence different expenditures, the consumer price index does not necessarily measure the cost of living for any particular individual. A homeowner, for example, would not be subject to rent increases or higher mortgage rates, so that part of the index would not apply.

It must be admitted that certain aspects of the consumer price index seem to invite a degree of skepticism. By a peculiar coincidence, the CPI always seems to look better just before an election. A close examination of some of the numbers, such as the cost of food, often makes us wonder where the evaluators are buying their groceries. Also the consumer price index uses accounting procedures that sometimes seem to include very arbitrary decisions. For example, the additional $1000 paid for emission controls and safety features on a new car do not appear in the consumer price index because these are regarded as improvements John Q. has gratefully made for humanity's sake.

If we put aside any questions as to the accuracy of the consumer price index and assume that it provides an approximate estimate of the cost of living for a typical individual in the United States, then what are the consequences of the double-digit inflation we have been having? An inflation rate of 12 percent a year means a doubling time of about six years; $1 today will be worth 50 cents in six years, 25 cents in twelve years, and only 12.5 cents in eighteen years. Suppose a parent decides to save money for a child's college expenses, which today are over $5,000 per year at a medium-priced institution. Assuming continuous inflation, in six years over $10,000 will be needed per year, in twelve years over $20,000, and in eighteen years more than $40,000 may be needed for just one year of college expenses. Multiply this by even two children, and it is easy to understand the hopeless feelings of many parents.

Similar problems face an individual who retires at age 65 with a fixed income of, say, $10,000 a year to supplement Social Security. A man can look forward to about fourteen years of additional life, at which time his retirement income will have been reduced to the equivalent of less than $1900 per year. A woman can look forward to about eighteen years of additional life, at which time her retirement income will have been reduced to the equivalent of about $1100 per year. Inflation is one of the chief worries of retired people, and it is not surprising that many of them decide to travel and enjoy themselves today rather than leaving their money to depreciate in a savings account.

Inflation can also be very hard on young people who want to establish their own home. Many young married couples have tried to save money for a down payment on a house, only to find that inflation was increasing the cost of housing so rapidly that the required down payment was growing faster than their savings. *The New York Times* recently included an interesting article about high school students in the very wealthy suburb of Scarsdale, New York. Some of these students were actually developing emotional problems because of concern that they would never be able to provide for themselves the affluent life of their parents' generation.

Most authorities are agreed that persistent inflation of this kind will sooner or later undermine the foundation of our democratic societies. A society where one cannot plan for the future is no longer a free society—its people literally have to live from hand to mouth, like savages. Much of today's suspicion and distrust of all forms of government is the direct result of their elected officials' apparent inability to stop inflation in spite of continual promises to "do something" about it. Inflation causes an income redistribution between those who can cope with it and those who cannot. It destroys the goal of saving (in fact, it penalizes saving) and encourages borrowing, as witness our "credit card economy." Inevitably, it increases tension among various levels of society and causes some, in desperation, to call for some type of totalitarian or fascist movement to "throw out the establishment" and start all over. Austrian author Stefan Zweig once commented, "Nothing made the German people so embittered, so raging with hatred, so ripe for Hitler, as the inflation." Dr. Arthur Burns, economist and former head of the Federal Reserve Board, recently told the Senate Banking Committee that inflation has brought the United States to a "crisis unmatched in its dangers since the great depression of the 1930s." *It is no exaggeration to say that, unless double-digit inflation can be stopped, the entire political structure of the free world is in jeopardy.*

Why We Haven't Stopped Inflation

Why are the necessary steps not taken to halt inflation? The plain truth of the matter is that, despite their protestations to the contrary, most of our government officials have found inflation to their advantage. To begin with, their own salaries are usually indexed to the consumer price index, so that they personally have nothing to lose from inflation. As wages and prices rise, tax income also rises and brings in more for the government to spend; this permits them to promise all kinds of benefits to special interest groups who will vote their way while the cost is covered in future years with depreciated dollars. In addition, inflation tends to make the national debt, now approaching a trillion dollars (that is, $1,000,000,000,000!), look a bit smaller. A yearly inflation of 12 percent, for instance, appears to justify a budget deficit of $50 billion since there is an even greater loss in the value of the inflated dollars. In the now classic words of President Franklin Roosevelt, "We only owe it to each other."

Particularly insidious is the effect of inflation on our income tax, sometimes referred to as the "invisible tax" or "taxflation." Suppose a 15 percent wage increase is granted to compensate for inflation. Every additional dollar is now subject to tax at the employee's highest tax bracket. After taxes, this 15 percent raise may well be reduced to a 10 percent "take home" raise which does not even equal the inflation rate.

In addition to those on the active and retired government payroll, there are other groups, such as organized labor and industry, who have been able to transfer the cost of inflation to others who are less favorably placed. How often is the following scenario repeated? A militant labor union in a critical industry announces that they are going to demand a generous "catch-up" wage increase. Industrial representatives say that the proposed increase will be strongly resisted as excessive and unjustified. Bitter negotiations start, a strike is threatened, negotiations break down, and a government mediator enters the arena. Miraculously, a last-minute agreement is reached which inevitably "meets the guidelines." The mediator leaves (a hero), the union gets almost everything they asked for, and the industry promptly raises prices—with Mary and John Q. Public picking up the tab once again.

Another factor that has encouraged inflation in the United States is the painful memory some people have of the serious depression of the 1930s. The steps needed to control inflation are, of course, *deflationary* by nature and are seen as threatening to trigger a recession or depression. This has tended to inhibit some government officials from taking action strong enough to control today's inflation. It is interesting to note that the situation in Europe today is quite the reverse; memories there are of the disastrous German inflation of the 1920s, when a bushel-basket of money was needed to buy a loaf of bread that had cost 1 mark in 1914. It is easy to understand why many of the Europeans have been much more cautious in their financial matters, and inflation is much less of a problem in nations such as Switzerland and West Germany.

As we will point out later in this chapter, putting the blame on our growing oil imports or on our weakening balance of trade in general cannot be justified by the facts. The chief reason we have such a serious inflation in the United States is the attitude of our people. Since World War II, we have made no distinction between material growth and growth in the amount of money available to us. An entire generation has been reared in expectation of continuous increases in salary. For many postwar years, these increased salaries were matched by higher productivity that acted to reduce labor costs. Today, when this is no longer true, our people continue to press for more money when it is not warranted by increased production. Industries raise prices to pay for higher wages and higher costs of capital equipment; this causes a demand for still higher wages, and a spiral is started. Many people have been fooled because it has the appearance of growth, but inflation is actually an expansion of money and credit relative to available goods and services and can only result in a continuously rising level of prices. Eventually fewer and fewer people are able to keep up with the pace, prices continue to rise, and the entire framework of a capitalistic society is undermined.

To make the situation even more serious, present government tax policy rewards the individual who spends or borrows money and penalizes the one who tries to save. When a person buys a piece of property with a modest down payment, the mortgage will be paid off with dollars cheapened more and more by inflation, and all the expenses for taxes and interest are deductible from his or her IRS return. When money is put in the bank, even the low interest received is subject to federal taxation, and there is no way to avoid loss.

Control of Inflation

Figure 27 shows how the United States national debt has increased over the past twenty years. Between 1960 and 1966, it was growing at slightly over 2 percent a year. By the early 1970s (when President Johnson's motto during the Vietnam War was "guns *and* butter"), it began increasing at a faster rate. Since 1974, it has risen consistently at a rate of about 9 percent a year. The budget deficit of $59 billion for the 1980 fiscal year was the second largest in our history.

If the graph for the national debt is placed on top of the graph for the consumer price index (Figure 26), it is found that the two curves are almost identical. This is no mere coincidence. If an individual runs up a debt, expenses

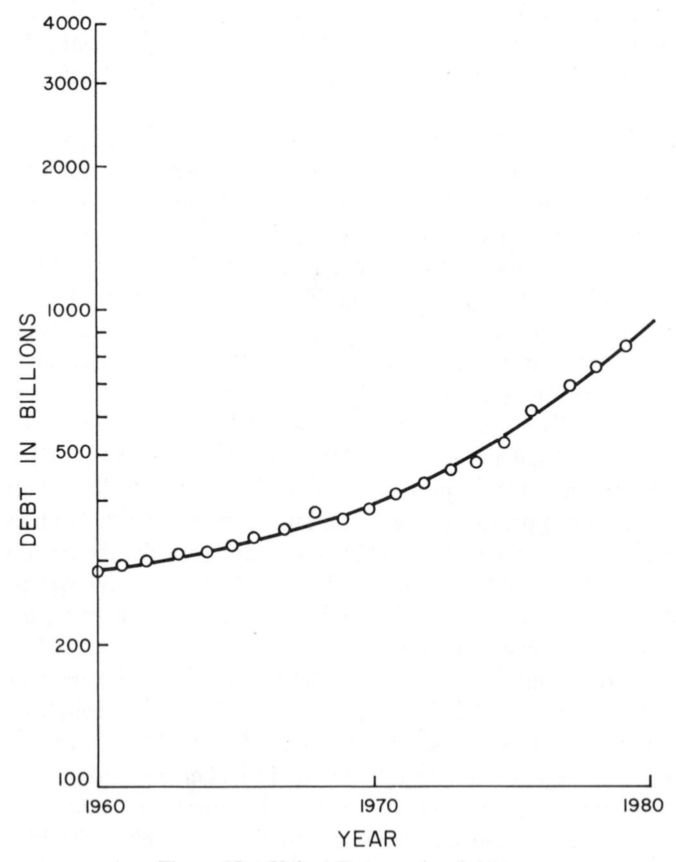

Figure 27. United States national debt.

must be cut down until the debt is paid. When a national government runs up a debt, however, it merely prints additional money and issues more bonds; like a "watered stock," this increases the total amount of money, cheapens its value, and results in inflation. With our national debt increasing at a rate of 9 percent annually and heading for the trillion dollar level in 1982, it is not surprising that the consumer price index shows a corresponding increase. *The only logical first step in controlling inflation, then, is to insist that our government exercise the same fiscal responsibility and restraint we all have to adopt in our personal lives; simply put, we have to live within our means.*

A second means of controlling inflation is to reduce the *money supply,* the total amount available for spending. The most common measure of our money supply is the total amount of all coin and paper money in circulation plus all *demand deposits* (all checking accounts and passbook savings accounts payable on demand). This count of the money supply is designated by the Federal Reserve System as *indicator M-1.* From Figure 28, it can be seen that the money supply has been increasing continuously for the past twenty years. During the early 1960s, it was growing at just under 4 percent a year; since 1975, it has grown at about 7 percent a year.

The United States Federal Reserve Board has the power to control the money supply by changing the interest rate on money made available to member banks. If the rate is decreased, banks can loan money to business and consumers at a lower rate, thereby expanding the money supply and the economy by encouraging more borrowing; if the interest rate is raised, the opposite occurs and there is a decrease in the money supply.

During the past twenty years, attempts have been made to control the money supply by lowering interest rates during business recessions to "pump up" the economy and by raising interest rates during severe inflationary periods to "cool" the economy. It has mostly been a case of too little and too late. As we have seen, it has never been good politics to attempt to control inflation. The fact that the money supply has increased continuously since 1960, in spite of the much publicized inflation of recent years, indicates quite clearly the refusal of a series of administrations to take effective action. While, admittedly, the situation is complex (many economists today are even downgrading the importance of the money supply), it is hard to believe that a growth of 7 percent or more per year in the supply of American money can do anything but cause more cheapening of the dollar and more inflation.

Another method suggested from time to time as a means of controlling inflation is to institute some form of price and wage control by government edict. This was used fairly effectively during World War II, but very rapid inflation occurred immediately after the controls were removed. Many years later, President Nixon also instituted price and wage controls for a time, but these were not very effective and were again followed by very rapid inflation.

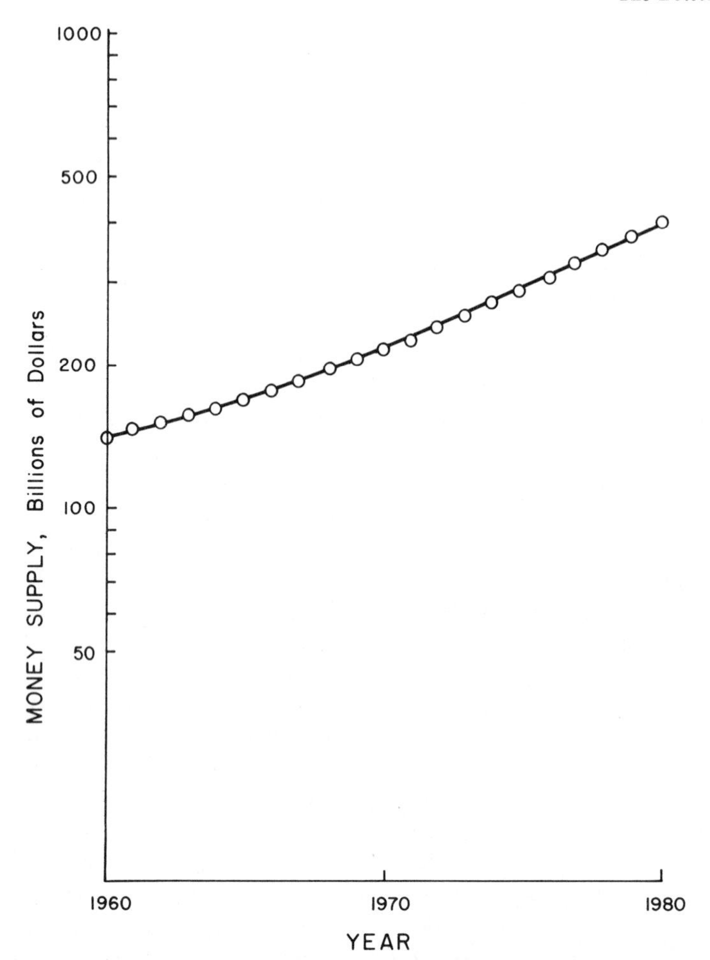

Figure 28. United States money supply.

President Carter's recent guidelines—enforced only by "jawboning"—were almost completely ineffective. Almost everyone agrees today that, in a capitalistic society such as ours, price and wage controls simply will not work.

Again, when all is said and done, the main reason for continuing rapid inflation in the United States is the jet-lag attitude of so many of our people. From 1945 until the 1973–74 Arab oil embargo, we were living in a culture dedicated to the principle of growth. Many of us still look on an inflated money supply as representing a better life, without realizing that increased prices can more than wipe out any increase in the dollars we receive. When we look at a nation such as Switzerland, which has been able to maintain stable prices in spite of the need to import practically all of its fossil fuels, we see clearly that this stability has been achieved because the people are frugal and conservative by nature and have insisted on these characteristics in their elected officials. When and if the people of the United States make it clear to our elected officials that failure to control inflation is losing them more votes than they gain by irresponsible spending to appease certain pressure groups, the first genuine move to control inflation will have been made.

PRODUCTIVITY

Another factor that has a strong influence on the cost of living is the efficiency of our productive system. If goods and services can be produced with high efficiency and low labor costs, this obviously makes it possible to sell them at a low price. This productive efficiency is measured in terms of *productivity,* namely the gross national product output adjusted for inflation and divided by the total worker hours of all people producing this particular output. The Bureau of Labor Statistics issues quarterly reports of productivity in various industries. Of particular interest is the productivity for all persons engaged in manufacturing, since this strongly affects the cost of our consumer goods.

For the twenty-year period from the end of World War II until 1965, productivity increased at an average rate of 3 percent a year (see Figure 29). This meant that each year an employer could look forward to a 3 percent decrease in his labor costs. It permitted employers to raise wages without raising prices: employers were happy because their costs remained the same; employees were happy because the additional money would buy more things. It was a twenty-year honeymoon. Furthermore, it did not take long for workers to feel that they were *entitled* to this ever-increasing disposable income, which effectively represented more pay for less work.

In 1965 the situation began to change. From 1965 until the Arab oil embargo of 1973, productivity increased at an average rate of about 2 percent a year. In 1974, following the embargo, there was a 5 percent *decrease* in productivity. It bounced back the following year, but since 1975 there has been very little increase in productivity and at times it has actually decreased. In the second quarter of 1980, productivity in the private business sector declined by 3 percent as a result of a 12.5 percent decline in output coupled with a 9.7 percent decline in working hours. Our entire productive system is in bad trouble.

The rapid increase in the consumer price index that started in 1965 was at least partially caused by the lower gains in productivity. No longer could increased productivity soak up wage increases; part of the cost of higher wages

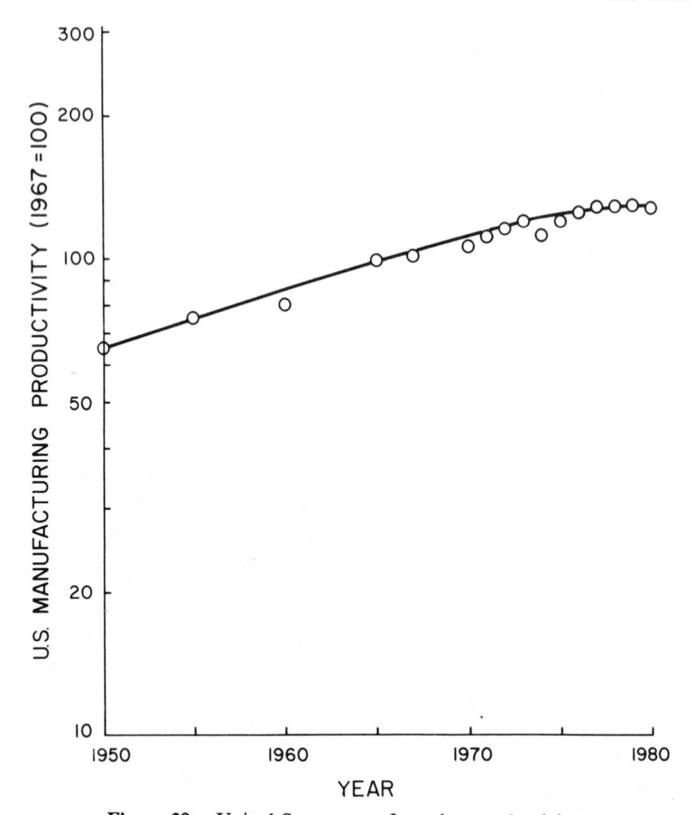

Figure 29. United States manufacturing productivity.

now had to come from higher prices. This caused workers to demand still greater increases, and the inflationary cycle became worse. *Since 1975, practically all increases in wages have been immediately reflected by corresponding increases in prices.*

What was the reason for the slowing down of productivity starting in 1965? It is probably no coincidence that 1965 marked the stepping up of the Vietnam War, resulting in the start of a rapid rise in the national debt and the resultant start of our serious inflationary period. Improved productivity comes from better ways of doing things and from the installation of labor-saving machines which enable a worker to produce more in a given time. (For this reason, productivity is said to be *capital-intensive*—that is, a business must have sufficient profits to pay for research and development and to buy the new labor-saving machines.)

From about 1965, the rapid increase in our national debt caused the federal government to take a greater fraction of the money supply. Bolstered by easy credit, homeowners and consumers also demanded ever-increasing amounts of money from our financial institutions, and there was not much left for private industry. Once the process got started, it automatically became worse. Consumers began to spend their money rather than to leave it in the bank where it would be available for industrial loans, more and more money went for interest on the increasing national debt, and interest rates began to rise. Now when industry had to replace a piece of equipment that was worn out, the money was hard to get, the interest rate was high, and (because of inflation) the new equipment cost two or three times that of the original. Under these conditions, industry had to cut back on capital improvements, research and development funds were reduced, and productivity suffered.

Today, labor unions are beginning to understand that it is essential they start to cooperate with management: to eliminate featherbedding, to find ways to improve productivity, and to avoid highly inflationary wage increases. Refusal to face these facts has sometimes been disastrous. For example, not long ago the Youngstown Sheet and Tube Plant in Youngstown, Ohio, was experiencing strong competition from imported products. It was estimated that their labor costs could have been reduced 21 percent merely by changing union seniority rules and reducing the plant manning levels that had been negotiated in contracts. When these changes were not made, the plant was permanently closed.

It is unfair, however, to blame the working man and woman for all our troubles. Only about 20 percent of today's labor force is in production, and the actual number of blue collar workers has been steadily decreasing in recent years. The other 80 percent of the work force consists of professionals, managers, technicians, sales, clerical, and construction people. The growth of bureaucracy in recent years has not been confined exclusively to the governmental sector. Managers, consultants, efficiency experts, attorneys, accountants, and a host of other occupations have multiplied throughout industry. All such people add to the cost of doing business but make no direct contribution to production.

There is also no question that government regulations have a negative impact on productivity. Burdensome tax laws, high environmental protection costs, and excessive regulation of industry drain off resources that could be used to improve productivity. Each year tens of thousands of pages of fine print in the *Federal Register* list all new government rulings and regulations, with some 2000 of them being legally binding. General Motors requires some 20,000 employees just to do federal paperwork at an annual cost of more than $1 billion. For a pharmaceutical manufacturer to obtain approval of a new drug requires preparation of enough documents to fill a small truck. In ten years there has been a six-fold increase in the budgets of some fifty-eight federal regulatory agencies, and it is estimated that new government health, safety, and environmental standards cost industry $100 to $150 billion a year.

Well-meaning government attempts to correct social problems also impact on productivity. Equal employment opportunities have added large numbers of women, blacks, and Hispanics to the work force. Largely untrained, these new workers often require considerable time to develop the necessary skills—and productivity suffers.

Management itself must also accept considerable blame for our productivity problems. In the United States, a corporate executive is judged almost entirely on the immediate results that are obtained, and his quarter-million dollar bonus depends on the "bottom line" in next year's financial report. Suggest an idea that may pay off in three years, and the answer is, "Forget it! Chances are I won't even be here in three years." Under these conditions, all actions tend to be based on obtaining immediate results, even if they may be harmful in the long run.

Can We Improve Productivity?

Perhaps the most serious aspect of the stationary United States productivity is that we are rapidly losing our ability to compete in the international marketplace. In the 1968–78 ten-year period, productivity increased in Japan by 89 percent, in West Germany by 64 percent, in France by 62 percent, and in Italy by 60 percent. The corresponding increase in the United States was 24 percent. Major industrial nations are rapidly overtaking and even passing us; they are producing goods at lower cost and underselling us in worldwide markets. One of every 6 tons of steel used in the United States now comes from foreign mills. Practically all shoes sold in the United States are foreign-made. In the 1960s there were twenty-five American-owned television manufacturers; today this is down to six. Hondas, Toyotas, Volkswagons, Volvos, and other foreign-made automobiles are taking over our highways. We are using more and more German beer, French wine, Scandinavian furniture, clothing from Taiwan, Korea, and Hong Kong; the list is almost endless.

Why has productivity continued to increase in other industrial nations, even those almost 100 percent dependent on imported oil? *In nations such as West Germany and Japan, workers and management see themselves as being in the same boat, and they prefer to stay afloat together rather than to sink together.* The attitude of the workers tends to be different from that in the United States; they have more pride in what they do, they consider themselves to be part of a growing nation, and this is reflected in their work. Management is perfectly willing to take a longer view of technology and promote research that may pay off in five or ten years. Privately funded research and development represents a much greater fraction of the gross national product in Japan, West Germany,

Switzerland, and the Netherlands. Industrial concerns work closely with universities to develop new ideas and to give an "industrial flavor" to academic studies. Government regulations are more reasonable, and are not applied with complete indifference as to cost or energy consumption.

Another important factor in these nations is the attitude of the public toward industry. People in general have more respect for private industry and feel more confident about the future. They are more cautious with their money, and this is reflected in the higher rate of savings that makes more money available to industry. For example, the Japanese save 20 percent of their income, the French 18 percent, and the West Germans 15 percent; Americans today are saving under 5 percent.

During almost thirty years of increasing material wealth (that is, disposable income) in the United States, anyone who has had contact with the working world has seen a gradual deterioration of morale at every level, together with increased dissatisfaction and continual bellyaching. Real craftsmanship, pride in work output, and genuine job satisfaction are becoming a thing of the past while, at every level, we look for the easy buck, the easy way out. Government jobs, with their built-in security, automatic raises, and generous retirement plans, are highly sought after. Recruiters from major corporations find that students on campus are far more concerned with stability than with opportunity. Top-drawer executive positions today must offer the proper "perks"—memberships in exclusive clubs, a chauffeur-driven limousine, stock options, perhaps a company-supported airplane and yacht—as well as a carpeted executive suite.

The development of new products and processes is an important aspect of productivity which many people do not seem to understand today. In 1979, the fraction of total corporate income devoted to research and development in the United States was at its lowest level in twenty years. Even worse, the entire nature of research and development has changed. In the past, its chief purpose was to develop improved products and processes and to find more efficient means of production. Today's research and development is increasingly *defensive* in nature, as it becomes more and more involved in finding methods to control pollution, obtaining data for environmental impact studies, and testing products to protect against potential liability suits. This means that the relatively small fraction of our expenditures that goes into research and development does little to increase industrial productivity.

In a resource-limited economy, any improvement in our material affluence will have to come from increased productivity. A complete change of attitude will be required. We must be willing, first of all, to conserve. We must recognize that the increased cost of our energy and raw materials will have to be met by cutting down somewhere else. Workers and management will have to cooperate and learn to take pride in such achievements as reduced energy consumption, reduced pollution, greater safety on the job, and more durable products. We must all realize that, even though our period of rapid growth is at an end, things are still pretty good in this country.

Many of our health, safety, and environmental regulations were passed before the energy crisis and our decline in productivity essentially changed the ballgame. These must be carefully reexamined with respect to cost versus benefits, required consumption of energy and materials, and their impact on productivity and inflation.

Among economists today, there are some with the opinion that productivity no longer has any real significance in view of the continuing growth of service industries. Others feel we should redefine productivity to include such things as energy and raw material consumption, increased employment opportunities, and credits for reduced pollution and for better health and safety of the work force. There is some validity to both of these approaches, but, regardless of how productivity is defined, our situation with respect to international markets is unchanged. The fact remains that we must export if we are to have money to pay for imported oil and other foreign goods. If other industrial nations continue to reduce production costs through higher productivity, the United States will find its worldwide markets becoming smaller and smaller until we have to enact import tariffs to protect domestic manufacturers from cheaper imported goods. Sooner or later, OPEC will probably refuse to take American dollars for their oil, and we will become an island surrounded by a protective tariff that has effectively cut off practically all exports and imports.

SCIENCE AND TECHNOLOGY

In the past 100 years, science and technology have brought us the highest standard of living the world has ever known. Indoor plumbing, central heating, pure food and water, clothing literally fit for a king, and the best medical care are taken for granted in the United States. Despite these advances, science and technology are looked on today with much suspicion by a large percentage of our population. Why is this so?

One answer may be that science and technology have become extremely complex, and the public naturally tends to feel uneasy about anything it does not really understand. This has been caused partially by the scientists themselves. World War II witnessed the development of entirely new and "way out" scientific achievements; atomic energy, synthetic rubber, antibiotics, television, and modern computers have radically changed our lives and have led to the creation of a scientific hierarchy which, in some cases, has practically dictated public policy. Completely ignoring any negative aspects, a rosy picture was painted of the brave new world of tomorrow that could be created if science were only given the opportunity. For a time, both government officials and the public accepted these statements without question.

In the 1960s, however, public uncertainty about the Vietnam War gradually led to a questioning of all aspects of business and government. It began to dawn on us that our highly touted new scientific and technological achievements were exacting a stiff price in the risks of nuclear power, the problem of radioactive wastes and increased pollution of all kinds, the computer-based invasion of our privacy, the extravagant consumption of our energy and raw materials, and our entrapment in the framework of a materialistic society. It became obvious that the solution of one problem could create others just as severe. We have seen, for example, how the lower death rate achieved by antibiotics and modern medicine has hastened the population explosion. Mass production and the modern assembly line have brought abundant cheap goods while acting at the same time to dehumanize the work place.

Another answer lies in the fact that the growth of science and technology also represents a typical *S*-curve. We are now going through a leveling-off period after a time of rapid geometric growth. This is clearly indicated by many signs: the general decline in industrial productivity, the current inability of research and development to turn up new ideas, the static number of new patents by United States corporations and individuals in spite of greatly increased research expenditures, the notable absence of new fields of scientific research, and the fact that much of today's technology is largely based on a mere extension of previous developments. While we shall undoubtedly see some spectacular new developments in certain new areas such as genetic manipulation of biological systems, areas such as chemistry and physics appear to have more or less reached a plateau.

The challenges facing science and technology today (improving energy efficiency, reducing pollution, developing substitute raw materials, and finding new sources of energy to meet future needs) are unfortunately not the sort of thing to "turn on" the average scientist or engineer. The kind of excitement and challenge in the development of atomic energy, for example, is entirely missing in the problem of reducing emissions from coal-fired power plants. In addition, many of the "small" problems are even harder to solve than the larger ones of the past, and there are the many frustrating government regulations to keep in mind. With most of the efforts of research and development directed toward the solving of these immediate problems, our rapid scientific growth has come to an end.

The Loss of Technology

In future years we shall increasingly face the actual *loss* of much of today's technology. In many industries this has already occurred because of environmental regulations regarding disposal of waste products. We noted earlier, for example, that the traditional Solvay process for the manufacture of synthetic soda ash has been essentially abandoned because of the problem of disposing of the waste product calcium chloride. Much of the pulp and paper industry as well has been forced to either shut down or change its processing operations to eliminate undesirable waste streams.

Our need to shift to consumer products that require less energy or use less critical raw materials will also result in a loss of technology. As we shift from aluminum cans to glass containers, much of the technology for the production and fabrication of aluminum will disappear. If we eliminate the chrome trim on our cars (probably a good idea since it is only for appearance and uses critical chromium), most of the technology of electroplating chromium will no longer be needed.

A particularly critical situation faces the four industrial companies that have been constructing our nuclear power plants: Westinghouse, General Electric, Combustion Engineering, and Babcock and Wilcox. With no new orders and with continued public opposition to nuclear power, it is unlikely that these concerns will all remain in the nuclear reactor business. Yet it is entirely possible that, in another ten or twenty years, we may have to start depending on more nuclear power or else do without new electric generation facilities; by that time, much of our nuclear power technology may be gone forever.

EMPLOYMENT AND UNEMPLOYMENT

Since World War II, it has been assumed in the United States that a growth in the economy (that is, the gross national product) of some 4 percent each year is required just to provide job opportunities for new workers entering the labor market. Does this mean that, as the economy levels off, we can expect unemployment to increase annually by an amount equal to 4 percent of our total labor force?

The situation to date has not been as bad as might be expected. New jobs have been created in recent years by the rapid growth of service industries and by the equally rapid growth of sports, fast-food outlets, and recreation industries. The ballooning of all types of government agencies, particularly those concerned with environmental, legal, and energy matters, has opened up many new jobs (funded, of course, by the taxpayers). There is, in addition, a silver lining for some in our static productivity, since many jobs continue to be done by humans rather than by more efficient machines.

On the negative side, however, with smaller amounts of discretionary income due to inflation, many of us are already beginning to cut back on vacation and leisure spending of all kinds. California's Proposition 13 and its successors across the land warn us that taxpayers may well increasingly oppose any further growth in government spending. The slowing down of productivity may prove to be temporary, as computers and industrial robots continue to take over the assembly lines. Equal employment opportunities are opening up the labor market to larger numbers of Hispanics, blacks, and women. Because of economic considerations as well as changing life styles, a greater percentage of American women are now working outside the home. Last but not least, large numbers of young adults have been seeking their first full-time employment as the effect of the postwar baby boom reached its peak.

It is also a fact that continuing increases in the cost of energy will literally price some activities out of business. This is already happening in the hotel and airline industries, where prices have risen so rapidly that travelers on expense accounts are taking over more and more and may eventually become almost the only ones on the road.

In an economy where we will sooner or later have to make choices as to the most effective use of the limited quantities of energy available, we can expect that a priority system of some kind will inevitably bring widespread unemployment in nonessential businesses, such as sports, recreation, travel, and leisure activities. On the other hand, the increasing cost of energy will in some ways serve to help the employment situation. As industry places greater emphasis on reducing energy consumption, the end result may be increased use of human power rather than installation of new energy-guzzling equipment.

The Workweek

One often hears the suggestion to create new job opportunities by reducing the number of hours in the workweek. Not too long ago, a 50- or 60-hour workweek was common. During World War II, essential workers were on a 48-hour week, and during our period of rapid postwar growth a 40-hour week was standard. With today's economy, it surely makes sense to question whether, in fact, a 30- or 35-hour workweek might be desirable. It would be better, for example, to have four people working a 30-hour week rather than three people working a 40-hour week with one unemployed.

The 40-hour workweek was established in 1938 by the Fair Labor Standards Act, which established at the same time a minimum wage of 30 cents an hour (how times have changed!) and required all employers to pay time and a half for overtime hours. While minimum pay has increased steadily since 1938, the 40-hour workweek and the overtime penalty have remained unchanged.

It is ironic to note that, in spite of the time-and-a-half penalty required for work in excess of 40 hours a week, many employees today are working a more or less regularly scheduled 48-hour week. This results from the fact that employers today must pay close to 50 percent of the hourly rate for insurance, pension, and fringe benefits. Since the overtime penalty is applied to the hourly pay alone, and there is no additional cost involved for fringe benefits, it is to their advantage to have employees work overtime rather than to hire additional workers. The employees are usually glad to earn the extra money; if it is necessary at some future date to cut back on plant production, the regular workers can return to a 40-hour week, and there are no problems of laying off workers, termination pay, or increased unemployment insurance. In 1979, during congressional hearings on a possible reduction of the workweek, Rudy Oswald, director of research for the AFL-CIO, said, "It is unconscionable for employers to be given this financial incentive to regularly schedule overtime work when large numbers are unemployed."

Most of the encouragement for a shorter workweek does come from organized labor, which undoubtedly hopes for fewer hours with no reduction in pay. Most of the opposition comes from industry. The Chamber of Commerce of the United States argues that any reduction of the workweek would be inflationary, would lower the workers' income, would reduce productivity, would disrupt collective bargaining, and could eventually lead to severe job shortages. Periodically, legislation is introduced in Congress to reduce the workweek, but it has never received much support. When Representative John Conyers of Michigan and twelve colleagues introduced a bill in 1979 to reduce the workweek to 35 hours, to pay double for all overtime, and to bar compulsory overtime assignments except in time of national emergency, their bill never even made it to the floor.

The question of fringe benefits is, of course, a strong deterrent to any reduction in the workweek and also acts to eliminate many part-time jobs. Employers are often required to provide equal fringe benefits for all employees, regardless of the number of hours they work; thus, two part-time employees each working a 20-hour week would cost twice as much in fringe benefits as one employee working a 40-hour week. It is obviously to an employer's advantage to keep the workweek as long as possible.

Although it appears inevitable that sooner or later a shorter workweek will become the law of the land, at least two rather serious difficulties could result. In spite of the high rate of unemployment, many employers today are faced with a tight job market for many positions, particularly those in the skilled trades. (Have you tried to find a plumber lately?) The number of new people entering the work force each year is beginning to drop as we pass the baby-boom effect, and some predictions indicate severe labor shortages in the late 1980s. It must also be remembered that a shorter workweek will not necessarily reduce unemployment of itself; it could easily lead merely to more moonlighting, with skilled workers holding two or even more jobs at one time. To be really effective, there would have to be some means to prevent an individual from holding two or more jobs.

The basic fact that everyone must keep in mind, however, is that our entire economy is increasingly resource-limited. It is simply a new way of life to which we must adjust. Material gains can come only from greater efficiency in the use of our energy and raw materials. It is essential that we handle these wisely and conservatively, and it is equally essential that the same care be taken in the handling of our labor resources.

"INCOME SECURITY"

Of real concern to most Americans today is the future of the many sprawling programs that Washington optimistically refers to as *income security.* In addition to the familiar Social Security, these include veterans' benefits, government and military pensions, railroad workers' pensions, medicare, medicaid, aid to families with dependent children, food stamps, and unemployment compensation. Today these take some 35 percent of the total federal budget as compared with 14 percent of a considerably smaller federal budget only twenty-five years ago (when some of these programs did not even exist). This is certainly an area where we will be faced with some very uncomfortable choices in the years ahead as we are caught between a growing percentage of nonproductive citizens and constant or decreasing resources. Figure 30 shows the rapid growth of the federal budget in the last 20 years.

Particularly troublesome is the increasing Social Security burden. Every day the number of Americans over 65 years of age increases by 1500 persons. In 1950, about 8 percent of our population was over age 65; by 1980 this had increased to 11 percent; it is expected to be over 12 percent by the end of the century, and some estimates predict that it will go as high as 20 percent during the first half of the twenty-first century when the postwar baby-boom generation reaches retirement age.

How It Works

With so many of us rightfully anxious about the future of our Social Security system, let us be clear about what Social Security is *not.* Social Security is *not* a bank where our payments are deposited in an account to be drawn on when needed. Social Security taxes simply go into a common pot where they are used to pay the benefits of those who are collecting today. Social Security was never intended to be the sole source of retirement income; it was designed to supplement individual retirement plans. It was not intended to be "fair"; its benefits were deliberately planned to provide a relatively greater income for those

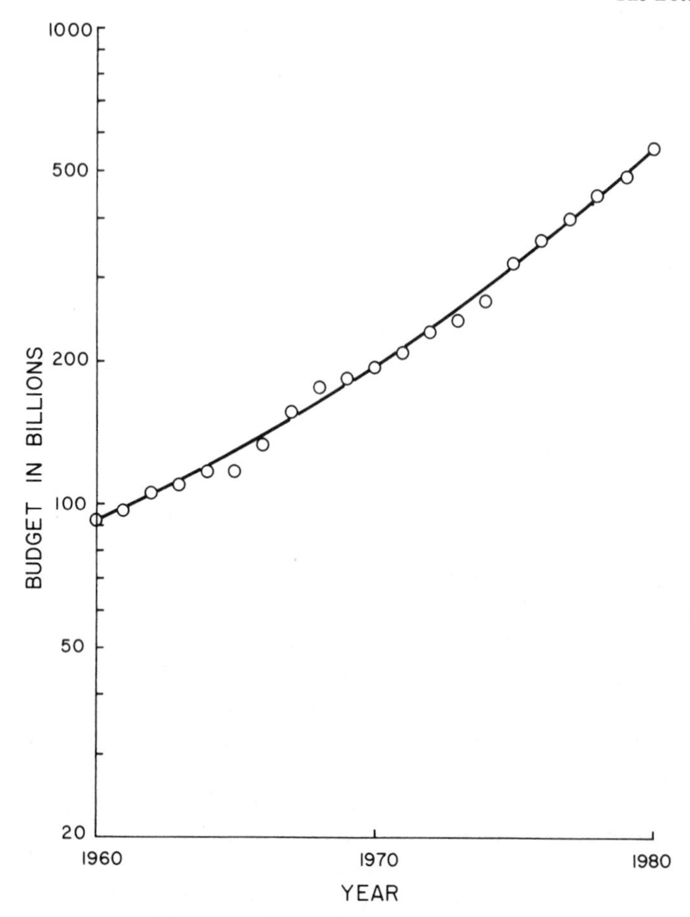

Figure 30. United States federal budget.

making small contributions, a fact that has proven quite lucrative for the retired government and military double-dippers.

It is frequently commented that the system is unfair to women or to bachelors or to some other particular group. The only answer to that is, "What individual are you talking about?" Consider the following.

1. Mary Jones and Sue Smith are both married and both "retired under Social Security." Mary contributed to Social Security for forty years as a result of her clerical job; Sue never worked outside her home and never paid a dollar in Social Security tax. Sue receives a Social Security retirement benefit equal to half of what her husband receives from Social Security. Since Mary had a fairly low-paying job, it is to her advantage now to also receive an amount equal to half of her husband's check, and she never actually receives a dime against the forty years of taxes she contributed personally to Social Security.

2. Tom Cohen and George Brown work at the same job and make the same payments to Social Security. Tom is married with six children while George is a bachelor. If Tom dies, his wife will receive Social Security payments for herself and the children until the youngest reaches the age of 18. If George dies, his estate will receive $255 for burial expenses.

3. Jane Green and Kathleen Stein are both retired and collect identical Social Security checks. Jane inherited an estate which pays her $20,000 a year from stocks and real estate, but this does not affect her Social Security payments. Kathleen is having a hard time making it on her Social Security retirement income; although she has been offered an interesting job, she turns it down because she knows that more than $5500 in yearly earnings will cause her to lose some or even all of her Social Security income.

There are three basic problems the Social Security system faces today. First, the entire concept of Social Security is like a chain-letter scheme which depends on a continuously increasing number entering the system to provide the benefits. In the 1930s, when President Franklin Roosevelt's "New Deal" was getting under way, there were a large number of workers and a small number of retirees; this has gradually changed until today every three workers support one Social Security recipient. In 1979, the Social Security fund *received* contributions of $99 billion from 114 million United States workers; that same year, it *paid* $101 billion in benefits to thirty-five million recipients. Over half the workers in America now pay more Social Security tax than income tax, and some authorities predict that the Social Security tax may have to be increased to 25 percent of taxable income by the end of the century if present trends continue unchanged.

The second basic problem with Social Security is that Congress has always been quick to increase benefits to win a few more votes, leaving future generations to worry about paying for the increased benefits. Benefits have been indexed to inflation, while the system's income usually lags way behind. In addition, various programs have been added to assist individuals who often never contributed to the system, such as medicare for the elderly. One of the fastest growing benefits of Social Security is disability insurance, where roughly one-quarter of the nation's five million disabled workers collect 80 percent or more of their former earnings. Since this money is not subject to tax, it represents a take-home salary greater than when they were employed. Not surprisingly, more and more disabled workers find it impossible to return to work.

The third basic problem of our Social Security system today is that people are living longer but retiring earlier (a majority do not wait until they reach 65). When Social Security was launched, the average age of death was about 62½, and it was assumed that most workers would not even live to collect under the system. Today's average worker lives into the 70s, and this will soon be extended by the longer lifespan of the increasing numbers of female workers. It has been seriously suggested that the retirement age should be increased to 68.

By 1977, the financial outlook for Social Security had become so bad that Congress was finally forced to act or see the whole system go broke. Small increases in the Social Security tax were approved for 1979 and 1980, with a larger increase to follow in 1981 (after the elections). In 1979, President Carter proposed a modest reduction in Social Security benefits such as eliminating the death benefit (for burial expenses) and college-student survivor benefits, reducing disability benefits, eliminating the retirement benefit minimum, and eliminating benefits for widows and widowers who were below retirement age. This was greeted with a loud horselaugh from Capital Hill, and his entire proposal was shot down.

Another suggestion often made is to partially fund Social Security benefits from ordinary federal tax revenues. It really makes no difference what pocket Social Security payments come from, the money has to be collected from the taxpayers in one way or another. It would be poor psychology to externally fund Social Security, however, since this would permit Congress to escape facing the realities of the system they have created.

It is clear that unless something can be done to keep our federal-income security programs from using a continuously growing percentage of the taxes of productive workers, the entire welfare system may collapse. More realistic ways will have to be found to phase people out of a productive working life, such as graduated retirement with a sliding scale of benefits. It has been suggested that more physically able retirees could themselves help in the care of the elderly in their own homes rather than placing them in the overcrowded nursing and convalescent homes. It has also been suggested that Social Security income be limited to those with demonstrated need, with the large number of relatively affluent elderly taking less to permit higher payments to the small fraction of genuinely poor persons.

The *unfunded* Social Security debt is now in excess of $4 trillion; that is, if no additional workers were to enter the system as of today, this amount has been promised by the government to workers already in the system. Even with new workers continually entering the system, taxes would have to be increased by some $14 billion annually just to continue benefits at the present level, and this amount will undoubtedly increase as inflation requires still higher payments. It is not surprising that a recent Harris poll found that 42 percent of the working people in America have "hardly any confidence at all" that they will ever collect the Social Security benefits they have earned.

FOREIGN AID AND FOREIGN TRADE

Almost everyone is agreed that shrinking resources must inevitably increase international friction as greater competition develops for continuously decreasing supplies of the world's critical raw materials. Today, the OPEC nations are reaping a rich harvest from their petroleum reserves; tomorrow, other nations will do likewise as other materials become hard to get. More and more, we can expect to see a barter system in use; for example, American wheat may be exchanged for Soviet platinum. Those nations with limited raw materials of their own, such as Japan, will find themselves at a growing disadvantage and will sooner or later have to reduce their population if they are to sustain a reasonable standard of living.

A second major effect will be in our attitude toward foreign aid. In the past, our economy could easily afford modest amounts of foreign aid, but with greater pressures on our limited raw materials in view, foreign aid will be possible only through real sacrifices on our part. There is also the sheer magnitude of the situation to consider: the rapidly growing undeveloped nations of the world now contain over three billion people, and even the relatively substantial resources of the United States today are quite inadequate to be of significant help to this mass of humanity.

The greatly increased cost of imported oil is currently having a disastrous effect on United States balance of trade, and this has inevitably raised doubts as to our ability to even continue foreign aid. Large amounts of energy are required to grow, harvest, and transport American wheat, corn, and rice, for example, and a good deal of this energy must come from imported oil. It is entirely possible that we shall sooner or later have to sell or barter *all* our excess food grains if we are to have foreign exchange to pay for our imported oil.

In Chapter 3 we have seen some of the unexpected problems that have resulted from foreign food aid; it is also true that some rather unpleasant surprises have been associated with our military aid to other nations. It is certainly high time we take a fresh look at what has been accomplished, what we are doing now, and what we hope to do in the future with the increasingly limited resources at our disposal. In restudying the nature of our foreign aid, we may find it necessary to establish new priorities and to place more and more emphasis on projects that involve genuine self-help.

situation to another putting out fires. This has caused considerable waste of our aid and has even raised serious doubts as to whether it was doing any good at all in some cases. Furthermore, it has resulted in some of the developing nations playing games with us, threatening to turn to the Eastern bloc of nations for a better deal—or, in the recent case of Pakistan, rejecting our aid offer of several hundred million dollars as "peanuts"!

THE CHOICE IS OURS

The entire capitalistic system of the United States was built on the Puritan work ethic, with stress on personal achievement, relentless pressure for advancement, and the acquisitive materialist drive for what was touted as the "Good Life." How can this be reconciled with a resource-limited world? Those "Type A" personalities who would have been so successful in the past may find great frustration in a society where there is no chance for economic expansion, little or no opportunity for research, and where many forms of creativity may actually be discouraged.

Whether or not free enterprise can even survive will depend on how long it takes us to accept the facts. If, for example, if becomes necessary to place controls on the production of certain finished goods, industrial concerns may be assigned or may voluntarily divide among themselves shares of the market (to more equitably distribute the limited production), production of energy-intensive materials (such as aluminum) may be eliminated altogether, and other products may be encouraged. Controls may also have to be placed on agriculture, with water, fuel, and fertilizer rationed. As energy shortages and higher raw material prices cut into profits, price and wage controls might even begin to look desirable.

Are we likely to see a change in our system of government, either by revolution or by gradual transition to a new form? It may well turn out that the problems ahead will require a government capable of rallying (and perhaps enforcing) compliance far more effectively than can be done in a democratic setting. There has certainly been nothing in the performance of Congress or the presidency during the past ten or twenty years to indicate the capability of handling the severe problems we now face. If it should come to the actual survival of the United States as a nation, a more authoritarian form of government may be unavoidable. It is a question of choice—ours and our children's.

A dictator is not necessarily identified by wild eyes, a loud voice, and a military uniform. Big Brother could be a well-spoken, clean-shaven Harvard graduate who wears suits from Brooks Brothers. It is even more likely that we will drift into a dictatorship run by a collection of Little Brothers. In fact, we are seeing today what may be the first signs of a more strongly regulated life ahead. We are already being told where our children must go to school, who our neighbors must be, what our working conditions must be, and what we can and cannot do with our property.

While the 1980 elections in the United States indicated very clearly that the average American wants strong and decisive leadership, they also indicated a desire for less, not more, government regulation. Things would have to be very bad indeed in America before there was any real public support for further government control of our lives. Does this mean that we are willing and able to control ourselves? This is the real challenge, and the choice is ours.

LESSONS FOR THE UNITED STATES

Inevitably at this point some will ask whether the answer might not lie in some form of socialism or communism. Thoughtful examination of the communist governments of Eastern Europe today shows that their "planned economies" have not been very successful (compare the standard of living in West Germany with that in East Germany, for example). In the resource-limited world of tomorrow, both socialism and communism will suffer from an added handicap: both are designed primarily to stimulate growth.

Great Britain, on the other hand, has "enjoyed" a stable-state standard of living for many years. During the decade of the 1970s, for example, its population was almost unchanged and its per capita income (corrected for inflation) increased at only about 1 percent a year. Great Britain has a strong political system, a homogeneous social culture, and relative freedom from problems of immigration or the disadvantaged. Here, if anywhere on earth, we would expect to find the model of a modern equilibrium society.

What has life been like in Great Britain during the past twenty or thirty years? Not long after World War II, Britain's voters turned to the Labour party (which was promising all things to all people at the time) and adopted a modified form of socialism: an expanded welfare system, free medical and dental care for everyone, subsidized housing, and nationalization of the steel industry as well as many other key industries. Unfortunately, it was soon found that the resulting business stagnation created just as much or more social and political stress as had the earlier untrammeled growth of private industry. There was strong resentment of unemployment and of the failure of living standards to rise. Without economic growth, it was impossible to spend more on social services such as education and housing. Social tensions were heightened and the government gradually lost the confidence of the people.

In 1970, the Labour party was voted out of office because the "prices bothered the housewives" and the public felt the Conservatives would be better at fighting inflation. Edward Heath took over as Prime Minister, but a national coal strike and other severe labor troubles brought down Heath's government in 1974. Not only had the Conservatives failed to solve inflation, but there was a disastrous trade balance, strong union demands for still higher wages, and almost every state-owned company was reporting heavy losses because the government barred them from raising prices. In October 1974, the Labour party under Harold Wilson took over the government with a new economic program that would emphasize industry first "to transform a declining economy into a high output high earnings economy." However Wilson's own party refused to support him and in 1979 the Labour party was again voted out of office and replaced by the Conservatives with Margaret Thatcher as Prime Minister. Her announced goals were to reduce taxes, scale down the welfare state, and reverse the decline of British society.

We also might take a look at Sweden, the most socialistic country in the free world today. In 1977, over 53 percent of the Swedish gross national product was taken by the government through taxes of one form or another (for comparison, it was 37 percent in Britain and 30 percent in the United States). All citizens receive subsidized medical and dental care, there is a yearly cash allowance for each child, and parents of a new baby can stay home for up to seven months at 90 percent of salary. When they are sick, workers receive 90 to 95 percent of their pay; not surprisingly, Sweden has a very high rate of absenteeism. It is almost impossible to fire or even lay off a worker. Industries receive government subsidies for job-seekers just out of school, for retraining workers they no longer need, and for producing and stockpiling goods they cannot sell.

Untouched by World War II, Sweden had an economic advantage over most of the nations of Europe. For many years there was continually increasing affluence in spite of wage and benefit costs close to the highest in the world. Interestingly enough, rising economic growth and affluence caused even more intensive insistence on equality, income redistribution and worker control of industry. A plan was proposed, and heartily endorsed by the labor unions, for labor to gradually acquire control of all Swedish industry.

Not completely satisfied with the way things were going, Swedish voters in 1976 ousted the Social Democrats after forty-four years in power and elected a nonsocialist coalition of the center, liberal, and moderate parties. In 1977 Sweden's economic honeymoon came to an abrupt end when the gross national product fell by 2.4 percent and the profits of all companies listed on the Stockholm stock exchange fell by 90 percent. The government was forced to take over the bankrupt shipbuilding industry and to heavily subsidize the ailing steel industry. It was recognized that the Swedish economy was very largely dependent on permanently weakened sectors such as paper, pulp, iron ore, steel and shipbuilding, all of which were facing increased global competition.

The problem Sweden faces today is that high labor and benefit costs coupled with only moderate increases in productivity make it very hard to compete in international markets. Even though wage increases have been restrained in recent years, there is considerable question as to the future. High taxes have removed any real stimulus for innovation, and executives over a ten-year period suffered a 20 to 30 percent decrease in real earnings because of taxes and inflation. Excessive bureaucracy has tended to stifle research and development and prevent the creation of new products and new industries. Perhaps their problems are best illustrated by the official of the Swedish Federation of Industries who had three placards in his office saying, in English translation: "Smile—you may be on radar," "Do it now—tomorrow there may be a law against it," and "There's no reason for it—it's just our policy."

How can the United States profit from the experiences of Britain and Sweden? These two democracies both have long-standing stable governments and homogeneous, ordered societies. Both have chosen to go down the path of socialism. Both are facing serious problems competing in international markets. Both now recognize that the only real hope to maintain their present standard of living, let alone improve it, is to stimulate private industry: modernize antiquated facilities, increase productivity, strengthen research and development, and thereby create new products and new technologies to open up new markets.

From the experience of Britain and Sweden, we can see that starting a little bit of socialism is like being a little bit pregnant. Once started, it almost inevitably continues to grow—and it is almost impossible to reverse the process. They also demonstrate clearly that *successful socialism requires a growing economy and thus cannot provide answers for a resource-limited economy.* Today, Sweden sees that its industrial future "requires research and development of new products," while Great Britain seeks to develop "a high-output, high-earnings economy." For each, this will be extremely difficult, because the motivation and initiative for creative research have been almost totally destroyed by the system.

It is also significant to note that both Britain and Sweden went into socialism with lofty ideals; in each case, the welfare state has gradually become what a Swedish housewife recently termed, "an overbureaucratized hierarchy, insensitive, willful and smug." In the United States today, many of us are already distrustful of government and resentful of bureaucracy. Our society is heterogeneous, and we do not like to take orders from government officials. The very modest moves we have made toward a welfare state in America have not been successful, but they have been extremely expensive. We all agree that there are many faults with our system of free enterprise, and that reasonable controls are always needed. The future resource-limited economy may require even more controls; however, if we go too far down the path of socialism, we, too, may wake up some morning to find that we have thrown the baby out with the bath water.

THE PURSUIT OF HAPPINESS

There is general agreement that the degree of happiness in any society has no relation to its material wealth. For example, the original Eskimos were considered to be one of the happiest races on the face of the earth despite severe hardship and almost total lack of material possessions. With one of the highest GNP's in the world, our happiness level in the United States today is certainly no greater than that of most of the undeveloped nations. Furthermore, Americans are no happier today than they were fifty or 100 years ago. Each generation takes for granted the standard of living it inherits. Today's children compare themselves to their friends; when the children next door visit Disney World, our children will feel deprived if they do not make the trip!

Several years ago, economist Richard Easterlin made a very interesting analysis of thirty surveys carried out in a total of nineteen developed and undeveloped countries to determine whether, in fact, money buys happiness. The conclusions were that, in all societies, an individual with a higher income than the average for the society as a whole will tend to be happier regardless of the absolute standard of living in the particular society, but *an increase in the overall standard of living does not increase the overall level of happiness.* This is the treadmill we in the United States have been on since the end of World War II. Each one of us has acted on the assumption that more money will bring more happiness, but when everyone acts on this assumption and all incomes increase by roughly the same amount, no one individual feels any better off. Like the Red Queen in *Alice in Wonderland,* each of us has been running hard for thirty years merely to stay in place!

Easterlin found that in every culture, happiness is largely determined by three personal considerations: economic position, family matters, and health —with economic position being the most frequently cited factor. Money cannot buy happiness, but evidently it helps! Considerations such as religion, politics, and general social conditions were mentioned much less frequently, perhaps by one person in ten. This is not surprising, since most of our time is spent at work and/or raising a family, and trying to deal with the many daily problems of marriage, housing, food preparation, education, and sickness. These problems are faced by people of all cultures and thus are the things that pretty much control our level of happiness.

What effect, then, can we expect a resource-limited economy to have on our overall level of happiness? If all of us suffer about the same decrease in material wealth, there should be no essential change in our happiness. If we can find a way to increase individual productivity ("a day's work for a day's pay") and learn to appreciate quality over quantity of output, job satisfaction may actually increase and offset our loss of material wealth.

There is always danger that a falling standard of living will increase social tensions among the various levels of society. In the past, many minority groups have been able to slowly work their way into the mainstream during periods of economic expansion of the system; such opportunities will be limited in a stationary or possibly a contracting work force. As we have seen, another likely consequence of a stable economy will be a decrease in average working hours and a corresponding increase in leisure time. Unfortunately, most Americans have never seemed able to use leisure time either wisely or gracefully. Weekends are filled with frantic activity—boating, camping, running, automobile trips, anything to escape from home; on Monday, it's back to work to rest up. In the spring of 1979 when there were again "gas" lines on the East Coast, a Long Island newspaper felt called upon to warn housebound weekenders not to mix tranquilizers, alcohol, and sleeping pills—evidently the only way some people could stand life at home for an entire weekend.

The initial adjustment to a resource-limited economy will be uncomfortable for many people, and it is quite likely that there will be a short-term decline in our overall happiness level. If we can redefine our values, particularly if we can learn to really *enjoy* leisure rather than excape from it by any means possible, it is probable that we Americans will eventually shift to a higher level of happiness and satisfaction than we have ever known.

"Bang the bell, Jack, I'm on the bus!"

The primary social problem facing the developed world has been best described by James Reid of Glasgow University, Scotland. It is simply that so many people today feel alienated by society, victims of economic forces beyond their control. Some degree of alienation has always been present, of course, but it is more widespread throughout all of society than ever before. It is probably also the most difficult problem that must be overcome in our adjustment to a resource-limited economy.

Many individuals do not understand what is bothering them, but it is expressed in their social attitudes and behavior. Alienation is most obvious in the senseless vandalism and antisocial behavior of young people and of all those who "cop-out" of society by seeking escape from reality through alcohol or drugs. Alienation is present even among those of us who appear to be adjusting to the world, as is witnessed by an increasing lack of confidence in most of our institutions. Even with Watergate several years behind us, a recent poll in the United States found that public confidence in Congress, big business, the news media, and many professions has sunk to a new low. Somewhat less obvious is the increasingly common form shown by individuals who have become literally alienated from their fellow humans—materialistic, self-centered, and unable to emotionally "get close" to anyone. Their motto is "take care of number one," or as they say in London, England, "Bang the bell, Jack, I'm on the bus!"

The truth of the matter is that we are, more than ever before, victims of economic forces beyond our control. Our standard of living depends on the whims of the OPEC nations. Our daily lives are increasingly affected by a faceless bureaucracy. Even our elected officials are so entrapped by the system they have created that they all behave pretty much alike.

This is no time for butting our heads against a brick wall. We must recognize our jet-lag attitudes and our alienation for what they are. It does no good to lash out at the system or at each other, to feel that we have somehow been cheated of our "rights." Americans can do something about energy problems, we can do something about inflation, and we can do something about employment—provided enough of us make up our minds what we really want! When and if our elected officials get this message from us and find out that it pays to be completely honest about the problems ahead, the adjustments will not be as painful as might be expected.

Adjustment will be particularly hard for those between the ages of about 20 and 40, who grew up during the period of incredible postwar growth in the United States. Theirs is a generation facing two severe psychological problems. First, because most of them accepted the material growth they experienced or saw all around them during their formative years as a way of life to which they were entitled, increasing numbers are now bitter and feeling "cheated" as they recognize that they will probably never be able to afford the excess of material things they have been conditioned to expect. The second problem for the generation results from the failure of their parents to pass along any real set of values. Growing up during the "permissive" postwar years, these young people (today's parents) had great confidence in themselves and strongly rejected any type of parental or other authority. Sexual freedom, drugs, rock music, and unconventional behavior characterized much of this period after surfacing explosively during the demonstration decade of the 1960s.

Now in its child-bearing years, this generation is suddenly faced with the responsibility of establishing a sense of values for their own children. In view of their background, it will be very hard for many of them to convey standards to help the next generation cope with a future of static or possibly declining material wealth. Yet it is essential for today's children to realize that a limit to material growth does not necessarily mean a limit to happiness. A way must be found to get the message across: many other goals are well worth working for, and life is still good in the United States. Schools, churches, and other community groups may help, but the most important factor will always be the attitude of parents themselves. If enough of today's parents can find meaningful standards for *themselves,* there will be no need for anxiety about how today's children adjust to their future world.

Speaking from seventy-odd years of a rich and diversified life, our situation is perhaps best described by C. Northcote Parkinson, author, historian, journalist, humorist, and creator of the well-known Parkinson's laws. In *Who's Who in America,* he has written:

> We have to discard our idea of progress. Civilization traces a curve of achievement. When we are on the upward slope we imagine that we shall rise indefinitely. But the curve levels off and then the descent begins. This has happened to western civilization in the twentieth century and all who have been taught to believe in progress feel bewildered, cheated, and lost. We should understand the situation better if we realized that it all has happened before, not once but many times, and that what follows summer must always be autumn. There is much we can do in autumn but not until we have admitted to ourselves that autumn has come and that winter lies ahead.

Selected Bibliography

Appleman, Philip, *Thomas Robert Malthus—Text, Sources and Background Material,* Norton, New York, 1976.

Barnet, Richard, *The Lean Years,* Simon and Schuster, New York, 1980.

Bingham, Jonathan, *Shirt Sleeve Diplomacy,* John Day Co., New York, 1953.

Boulding, Kenneth E., *Collected Papers,* Colorado Associated University Press, Boulder, Colorado, 1973.

Boulding, Kenneth E., *The Meaning of the Twentieth Century,* Harper and Row, New York, 1964.

Boulding, Kenneth E. et al., *Economics of Pollution,* New York University Press, New York, 1971.

Boulding, Kenneth E., Michael Kammen and Seymour Lipset, *From Abundance to Scarcity,* Ohio State University Press, Columbus, 1978.

Brown, Lester R., *By Bread Alone,* Praeger Publishers, New York, 1974.

Carson, Rachel, *Silent Spring,* Houghton Miffin, Boston, 1962.

Connelly, Philip and Robert Perlman, *The Politics of Scarcity,* Oxford University Press, London, 1975.

Dahrendorf, Ralf, *The New Liberty,* Stanford University Press, Stanford, 1975.

Darwin, Charles Galton, *The Next Million Years,* Robert Hart-Davis, London, 1952.

De Golyer and Mac Naughton, *Twentieth Century Petroleum Statistics,* the authors, One Energy Square, Dallas, TX 75206.

Easterlin, Richard A., "Does Money Buy Happiness?," *The Public Interest,* Winter 1973.

Fritsch, Albert J., *The Contrasumers,* Praeger, New York, 1974.

Hardin, Garrett, *The Limits of Altruism,* Indiana University Press, Bloomington, 1977.

Hardin, Garrett, *Population, Evolution and Birth Control,* W. H. Freeman, San Francisco, 1969.

Heilbroner, Robert L., *An Inquiry into the Human Prospect,* Norton, New York, 1974.

Himes, Norman E., *Medical History of Contraception,* Williams and Wilkins, Baltimore, 1936.

Jordan, Henry A. et al., *Eating is Okay,* New American Library, New York, 1978.

Klein, Rudolf, "Growth and its Enemies," *Commentary,* June 1972.

Knowlton, Charles, *Fruits of Philosophy,* Peter Pauper Press, Mount Vernon, New York, reprinted 1937.

Lloyd, William F., *Lectures on Population, Value, Poor-Laws and Rent,* Augustus M. Kelley, New York, reprinted 1968.

London Sunday Times Insight Team, *The Yom Kippur War,* Doubleday, Garden City, New York 1974.

Mancke, Richard B., *Squeaking by,* Columbia University Press, New York, 1976.

Meadows, Donella H. et al., *The Limits to Growth,* Universe Books, New York, 1972.

Meek, Ronald L., ed., *Marx and Engels on Malthus,* International Publishers, New York, 1954.

National Research Council, *Energy in Transition 1985–2000,* W. H. Freeman, San Francisco, 1980.

Novick, David, *A World of Scarcities,* Wiley, New York, 1976.

Ordway, Samuel H., Jr., *Resources and the American Dream,* Ronald, New York, 1953.

Paddock, William and Paul, *Famine-1975!,* Little, Brown, Boston, 1967.

Parkinson, H. Northcote, "Thoughts on my life," *Who's Who in America,* 41st edition, Marquis Who's Who, Chicago, 1980.

Peccei, Aurelio, *The Human Quality,* Pergamon Press, New York, 1977.

Peterson, William, *Demography 8,* 25 (1971).

Place, Francis, *Illustrations and Proofs of the Principles of Population,* Houghton Mifflin, Boston, reprinted 1930.

President's Materials Policy Commission, *Resources for Freedom, Volume I,* U.S. Government Printing Office, Washington, June 1952.

Reid, James, "Bang the Bell Jack, I'm on the Bus," *The New York Times,* June 20, 1972.

Rosen, George, *A History of Public Health,* MD Publications, New York, 1958.

Rostow, W. W., *Getting from here to there,* McGraw-Hill, New York, 1978.

Schurr, Sam H. and Bruce C. Netschert, *Energy in the American Economy, 1850–1975,* Johns Hopkins Press, Baltimore, 1960.

Simon, William E., *A Time for Truth,* Reader's Digest Press, New York, 1978.

Singer, S. Fred, *The Changing Global Environment,* D. Reidel., Boston, 1975.

Smith, Adam, *The Wealth of Nations,* Dutton, New York, reprinted 1933.

Stobaugh, Robert and Daniel Yergin, *Energy Future,* Random House, New York, 1979.

Udall, Stewart, Charles Conconi, and David Osterhout, *The Energy Balloon,* McGraw-Hill, New York, 1974.

United States Department of Energy, *Monthly Energy Review,* Superintendent of Documents, Washington, DC,

Vogt, William, *People,* William Sloane Associates, New York, 1960.

Vogt, William, *Road to Survival,* William Sloane Associates, New York, 1948.

Walford, Cornelius, "The Famines of the World," *Royal Statistical Society Journal, 41,* 433–526, 1878.

Wrigley, E. A., *Population and History,* McGraw-Hill, New York, 1969.

Yannacone, Victor J., Jr., *Energy Crisis,* West Publishing, St. Paul, Minnesota, 1974.

Index